UPCYCLING RESEARCH

UPCYCLING RESEARCH

Global Perspectives on Product, Fashion and Theory

Edited by Kyungeun Sung and Richie Moalosi

BLOOMSBURY VISUAL ARTS
LONDON • NEW YORK • OXFORD • NEW DELHI • SYDNEY

BLOOMSBURY VISUAL ARTS
Bloomsbury Publishing Plc, 50 Bedford Square, London, WC1B 3DP, UK
Bloomsbury Publishing Inc, 1385 Broadway, New York, NY 10018, USA
Bloomsbury Publishing Ireland, 29 Earlsfort Terrace, Dublin 2, D02 AY28, Ireland

First published in Great Britain 2026

Cover design by Louise Dugdale

A catalogue record for this book is available from the British Library.

Library of Congress Cataloging-in-Publication Data
Names: International Upcycling Festival (2024 : De Montfort University) author | Sung, Kyungeun editor | Moalosi, Richie editor | De Montfort University host institution | the International Upcycling Research Network sponsor
Title: Upcycling research : global perspectives on product, fashion and theory / edited by Kyungeun Sung and Richie Moalosi.
Description: London ; New York ; Oxford ; New Delhi ; Sydney : Bloomsbury Visual Arts, 2026. | Research presented at the International Upcycling Festival 2024 at De Montfort University, UK sposored by the International Upcycling Research Network (IURN). | Includes bibliographical references and index.
Identifiers: LCCN 2025026508 | ISBN 9781350496316 hardback | ISBN 9781350524637 paperback | ISBN 9781350496323 pdf | ISBN 9781350496330 epub
Subjects: LCSH: Upcycling (Waste, etc.)–Research–Congresses | Industrial ecology–Congresses | Waste minimization–Congresses | New products–Environmental aspects–Congresses | Clothing trade–Environmental aspectsvCongresses | Sustainable engineering–Congresses | Circular economy–Congresses | LCGFT: Conference papers and proceedings | Conference papers and proceedings
Classification: LCC TD794.9 .I58 2024 | DDC 628.4/458—dc23/eng/20250905
LC record available at https://lccn.loc.gov/2025026508

ISBN: HB: 978-1-3504-9631-6
ePDF: 978-1-3504-9632-3
eBook: 978-1-3504-9633-0

Typeset by RefineCatch Limited, Bungay, Suffolk
Printed and bound in Great Britain

For product safety related questions contact productsafety@bloomsbury.com.

To find out more about our authors and books visit www.bloomsbury.com and sign up for our newsletters.

To all the incredible researchers, academics and practitioners working on upcycling, circular economy, and sustainable design, production and consumption around the world

CONTENTS

FIGURES

TABLES

CONTRIBUTORS

Amal Abuzeinab, an academic advisor and architect at De Montfort University's School of Art, Design, and Architecture (ADA), currently serves as the Institute Head of Research Students at the Institute of Architecture. Her interdisciplinary research focuses on building the business case for sustainability within the Architecture and Built Environment disciplines. Amal uses business model concepts and tools as innovative approaches to green transitions. These led to the creation of a practical management tool that aids green/sustainable transitions at the organizational level. Amal has completed various funded projects and has published over 20 peer-reviewed academic articles and book chapters.

Luca Alessandrini, a PhD candidate in Design at Politecnico di Milano in Italy, and conducts research on the intersection of design and industrial symbiosis to develop materials from organic waste, forming localised company networks for upcycling by-products. With an international background in design, including an MA/MSc from the Royal College of Art and Imperial College London, his expertise spans circular economy, sustainable design, and material research. Collaborating with various European companies and institutions, Luca applies his skills to a diverse range of sustainability-focused projects. His R&D work demonstrates his dedication to advancing circular economy practices and developing cutting-edge sustainable design solutions.

Joon Sang Baek is Professor and Director of the DESIS (Design for Social Innovation and Sustainability) Lab at Yonsei University in South Korea. His research area lies in the nexus of service/systems design, sustainability and social innovation. He is currently interested in complex sociotechnical systems, ecocentric design and education and child-nature interaction design. His recent publications include 'Unleashing the Potential of AcciMap in Identifying and Anticipating Systemic Failures: A Case Study of Child Abuse Prevention through Design' (2023) and 'Network Analysis of Complex Stakeholder Needs for Service Ecosystem Sustainability: A Case Study of South Korean ESCO industry case Study of South Korean ESCO Industry' (2022).

Lovish Bandwal is a Senior Product Designer at Swiggy India, a leading food delivery tech company. He holds a Master of Design from IIIT Jabalpur. Lovish specializes in upcycling and studying the behavioural impact of music. His notable achievements include winning the MIT Hackathon and filing a design patent, along with accolades from several international competitions. Lovish is dedicated to enhancing user experience through sustainable, economical, and scalable designs. Outside his professional work, he enjoys travelling and playing musical instruments, reflecting his passion for creativity and exploration.

Benjamin Bolaane is a professional civil engineer and Professor of Infrastructure Planning and Engineering at the University of Botswana. His main areas of engagement and research are sustainable infrastructure provision and, more specifically, waste, water, sanitation, transportation and road safety. He has worked on projects in waste management strategy and policy development. His research work is published in a wide range of academic journals focusing primarily on enabling equitable access to infrastructure, particularly for the downtrodden. Recently, his research work has opened a new frontier of sustainable housing provision in urban areas.

Chanjuan Chen is Associate Professor in the Department of Design at the University of North Texas in Denton, USA. Her areas of focus include sustainable design for apparel and design research. Her creative scholarship on upcycling and modularity has been recognized through several design awards as well as a jury and invited exhibitions nationally and internationally. Chanjuan's interest in upcycling includes exploring formalized design processes for repurposing apparel and textiles through creative practice-based research. Her commitment to sustainable design practice has also been implemented in her teaching through developing fashion-sustainable studio courses and projects on upcycling.

Kevin Cheung is an upcycling product designer from Hong Kong, who is dedicated to developing upcycling products and raising environmental awareness among the public. He believes that designers have the ability and duty to protect the environment to maintain the balance between the ecosystem, economy and society. His focus is on transforming waste into attractive and contemporary works of art and products.

Walter Chipambwa is Lecturer in the Department of Clothing and Textile Technology, School of Art and Design at Chinhoyi University of Technology in Zimbabwe. He is pursuing his PhD studies with the University of Botswana, and his interests are in sustainable design, design-led innovation, computer-aided design, design education and management. Walter's interest in upcycling started with his first project in 2019. He is currently working with the community and industry on issues of sustainability and upcycling so that all players, from manufacturers to consumers, are completely involved.

Chipo Chuma is Senior Lecturer in the Clothing and Textile and Clothing Technology department at Chinhoyi University, Zimbabwe. She received a Fulbright visiting research fellowship to study interdisciplinary research and methods in textiles and clothing production. Her research interests include sustainability, circular fashion and accessory consumption.

Dazel Empe has a Master's in Engineering Management from De Montfort University, and a Bachelor's degree in Engineering Management from De Montfort University. He is currently a trainee consultant with an engineering firm.

Claudia García Flórez is an environmental engineer and researcher from Colombia. She has been working on projects related to sustainability, circular economy, and GIS technology applications as a tool for decision-making as part of triple-impact projects in Colombia and Australia. She led an upcycling initiative that was granted into the best 500 socio-environmental projects in the Latinoamérica Verde Awards 2020. She is passionate about circularity applied and providing innovative solutions. Her main research interest resides in sustainability, nature-inspired design, regenerative systems, biotechnology, innovation, and social-human impact.

Yagnesh Gohil is a Senior Experience Designer at Cognizant who stands out for his natural empathy, punctuality, and unwavering optimism. His passion for Digital Design seamlessly intertwines innovation with sustainability, reflecting his commitment to making a difference. With a in Master of Design from IIITDM, India, Yagnesh blends academic excellence with real-world experience. His flair for research and patent ownership highlights his drive for impactful solutions. Yagnesh continues to shape user experiences that resonate deeply, and championing sustainability—a true testament to his dedication to creating a better, more humane world through design.

Trina Harlow is Director of Art Education Undergraduate Studies for the College of Visual Arts and Design and university supervisor for the College of Education at the University of North Texas in the USA. She is the author of Journey to Refuge: Understanding Refugees, Exploring Trauma, and Best Practices for Newcomers and Schools and the co-director of the full-length film *Refuge in the Heartland: Refugee Resettlement Documentary*. Dr Harlow holds a B.S. in Fashion Design, an MA in Art Education, and a PhD in Education. She serves on the National Board of Directors for the National Art Education Association.

Karen Marie Hasling is Associate Professor at Design School Kolding, Denmark, where she co-leads the Lab for Sustainability and Design. Coming from textiles technology and design engineering, her research focus is on the intersections between materials, design, sustainability and education. Furthermore, she heads the interdisciplinary programme, MA Design for Planet. She is the co-author of Sustainable Design Cards and Material Pathways, two card decks that explore materials in design's role in sustainable transition. She is currently leading several research projects on materials in design education, design competencies in sustainable transition and design for textile circularity.

Yufei Hu is a PhD student in the Faculty of Arts, Humanities, and Cultures at the University of Leeds. Her research interests are in sustainable fashion, creative upcycling, and emotional design. Her PhD research topic focuses on Chinese fashion upcycling practice and develops an emotional design guideline for fashion upcycling practitioners.

Eunsuk Hur is Lecturer in Fashion Innovation and Marketing and a research group leader for the Fashion Marketing and Management Group at the School of Design at the University of Leeds. She is a founder of the Rebirth Fashion community, supporting local circular fashion activities, including clothing exchange, repair, upcycling, and sustainable lifestyle. Her research interests cover bridging design innovation and marketing, ranging from upcycling, consumer experiences, circular fashion community activities, collaborative consumption and the roles of social media and technology in leveraging social innovation.

Patrick Isherwood is Lecturer in Solar Energy at the Centre for Renewable Energy Systems Technology, Loughborough University, UK. His research interests include the development of sustainable materials, devices and processing techniques for solar energy conversion and storage, solar resource modelling, and thermodynamic modelling of solar collectors. He has a particular interest in the development of environmentally benign techniques for materials synthesis and for end-of-life device processing. He has authored more than twenty peer-reviewed papers and conference proceedings and is frequently invited to review articles for journals in solar energy and related fields.

Alice Jarry is Associate Professor of Design and Computation Arts at Concordia University. She is Research Chair in Critical Practices in Materials and Materiality, Co-Director of Hexagram, and Director of Milieux Institute Speculative Life Biolab. As an artist-researcher, she specialises in site-specific works, art-science practices and socio-environmental design. Her research brings concerns about aesthetics and politics to bear critically upon material production and infrastructures. Focusing on residual matter and active biomaterials/composites for the built environment and the arts, Jarry examines how materiality – engaged in processes of transformation with site, technology, and communities – can provoke the emergence of resilient socio-environmental relations.

Gerald Kennedy is Lecturer in Graphic Design at the Department of Design and the Built Environment, School of Arts and Humanities, University of Huddersfield, UK. He has research interests and expertise in graphic design, web design, app design, creative coding, UI/UX design, motion graphics and photography. He has publications on design, UX, education, culture, AI, and HCI. So far, this has culminated in various published articles, guest speaker invitations, and presentations at conferences both in the UK and abroad.

Busayawan Lam is Reader of Design at Brunel Design School, College of Engineering, Design and Physical Sciences, Brunel University London. She was trained in industrial design and practised as a product designer in a small-to-medium-size exporting company in Thailand. She later obtained a research degree in design management. She specializes in strategic design and has many years of experience studying innovation management and recommending strategic design directions for a variety of organisations, ranging from an equipment producer for domestic general hospitals to a global electronics company. Her current research interests include co-design, community-led design, and design education.

Montida Lertnimanoradee is a doctoral researcher at the Design School, Brunel University London – BA in Industrial Design from Khon Kaen University and an MSc in International Fashion Marketing from Coventry University London. Montida's research interests include graphic design, branding, and upcycling. She has extensive experience as a design lecturer. She has worked on numerous projects, including branding and media creation for local entrepreneurs, designing for the Khon Kaen International Marathon, and developing upcycling products for Bangkok Hospital Khon Kaen. Her PhD research focuses on participatory design and upcycling practices to transform municipal solid waste into products that benefit local communities.

Paulson Letsholo is Lecturer in Industrial Design and Engineering Product Design at the Faculty of Engineering and Technology, Department of Industrial Design and Technology, University of Botswana. His research focuses on the areas of Design for sustainability, Design thinking process, Design education and Eco-entrepreneurship through upcycling for purposes of widening the boundaries of design intervention at the grassroots level to contribute positively to the pillars and principles of sustainable development such as; Selection of resources with low environmental impact, Design of products with low environmental impact; Product-Service System Design for eco-efficiency, and Design for social equity and cohesion.

Tony Lorance is a freelance mechanical designer and graduate of the MSc Product Design programme at De Montfort University (DMU), UK. He has a thorough understanding and experience of computer-aided design, engineering drawings, reverse engineering, and manufacturing processes in the automotive industry. In 2023, he worked on the NPD4CE (New Product Development for Circular Economy and Upcycling) project, funded by DMU HEIF (Higher Education Innovation Funding) and supervised by Dr Kyungeun Sung, Mik Pieniazek and Dr Abhishek Tiwary.

Richie Moalosi is Professor of Industrial Design at the Department of Industrial Design and Technology, Faculty of Engineering and Technology, University of Botswana. His research interests include design and sustainability, social innovation, design and culture, design research, design education, and additive manufacturing. He is a Co-Investigator of the Arts and Humanities Research Council (AHRC)–funded project International Upcycling Research Network.

Thatayaone Mosepedi is Lecturer in Industrial Design at the Department of Industrial Design and Technology, Faculty of Engineering and Technology, University of Botswana. He is a distinguished figure in the field of industrial design, with a focus on empathic product and service development guided by the African philosophy of Ubuntu. His work emphasizes the importance of empathy in design, enabling designers to make informed choices that resonate with users' needs. He has contributed to the advancement of industrial design practices, particularly in emerging economies, through his expertise in new product development, design research and cognition.

Edlight Mutungwe is a senior lecturer who led as Chairperson in the Department of Curriculum and Instruction, then Chairperson in the Department of Clothing and Textile Technology at Chinhoyi University of Technology, Zimbabwe. She has interests in social psychology, the social psychology of dress, the history of fashion and research methods for art and design. She has done research in waste management, where she received two grants that resulted in the development of a centralized bin collection model, research in curriculum issues, life-long learning, sustainable fashion and fashion exhibitions.

Sophia N. Njeru is Lecturer in Fashion Design at the Department of Fashion Design and Textile Technology, Kirinyaga University, Kenya. She is avid about sustainable fashion in higher education and industry pertaining to upcycling, specialized apparel, indigenous dress, and research. She has authored/co-authored numerous peer-reviewed journal articles, conference proceedings, and book chapters, facilitated postgraduate colloquiums, as well as reviewed conference papers, books and journal manuscripts. Her most recent publication is entitled *A Sustainable Fashion Safari* (2024). Njeru is a member of the International Upcycling Research Network, Learning Network on Sustainability and the Kenya Bureau of Standards readymade garments technical committee.

Isabel Ordóñez is an instructor and researcher at Elisava Barcelona School of Design and Engineering. She holds a bachelor's degree in industrial design from the University of Chile and both a Master's degree and a PhD in Sustainable Design at Chalmers University in Sweden. As a designer, she works primarily with furniture. In her teaching and research, she has engaged in a diverse range of sustainability projects, particularly focusing on resource recovery from waste. Her current research area is education for sustainable development, with the aim of facilitating material reuse through the establishment of Urban Resource Centres.

Hamilton Ortiz is a Design Doctorate Student in the Faculty of Architecture and Urbanism at São Paulo University, with undergraduate and Master's degrees in mechanical engineering. His research and expertise are focused on the development of productive processes for upcycling in Latin America through hands-on experience and practice. He also has more than 13 years of experience in energy transition research and consultancy and has published papers about thermodynamic assessments of industrial systems.

Patrick Pearce is a designer-maker of furniture and objects, design researcher and communication strategist based in Montréal, Canada. He holds a Master of Design from Concordia University, where his practice-based, primary and secondary source research-creation centred on improving the scalability and viability of upcycling projects and enterprises from a designer and enterprise perspective. This project includes a collection of furniture and objects created from repurposed manufacturing materials. Informed by experiences in consulting, design entrepreneurship and filmmaking, his research interests include repurposing, sustainable craft, designer-makers, design processes, circular economy business models, and the storytelling value of upcycled objects.

Eujin Pei is Reader in Additive Manufacturing and Associate Dean for the College of Engineering, Design and Physical Sciences at Brunel University London, UK. He is the Director of the BSc Product Design Engineering programme. He teaches New Product Development with a research focus on Design for Additive Manufacturing/3D Printing. Eujin is a Fellow of the Institution of Engineering Designers (FIED) and chairs the UK national committee (BSI AMT/8) and international standardization committee for Additive Manufacturing Data and Design (ISO TC261/WG4). Recent publications include the *Springer Handbook of Additive Manufacturing* (ISBN: 978-3-031-20751-8).

Mik Pieniazek is Senior Lecturer in Product Design at De Montfort University, UK. He has extensive experience as a professional product design practitioner and educator in multiple higher education institutions in the UK. His research interests include NPD processes, urban metabolism, sustainability, and collaborative processes.

Martin Racine is Graduate Program Director and Professor in the Department of Design and Computation Arts at Concordia University, Montréal, Canada. His research explores design for disassembly and 3D printing to extend the life of objects. In the Meta-Morphose project, his team explored how objects could change, adapt, and evolve over time, using participatory design to develop a second life for discarded or outdated objects. In DNA – The Future Life of Objects, his team advances the concept of biomimicry, suggesting that sustainable objects could be integrated into the natural world as living organisms.

Yaone Rapitsenyane is Lecturer in Sustainable Design and Head of the Department of Industrial Design and Technology, Faculty of Engineering and Technology, University of Botswana, Botswana. His research covers the areas of design education, design practice and sustainability, focusing on product service systems and service design. He has been involved in various collaborative projects investigating and exploring sustainable design in education and in practice, service design in hospitality and industrial design in the design and prototyping of medical devices and implants.

Millie Rees is a BA Product Design student at De Montfort University (DMU), UK. Millie completed the DMU Frontrunners Internship on Computer-Aided Design and Upcycling between 2021 and 2022, supervised by Dr Kyungeun Sung.

Anneloes Roelandschap is a co-founder and CEO of CHAKO – a green social enterprise based on the island of Zanzibar in Tanzania focused on producing a range of upcycled products utilizing waste generated on the island and creating a circular economy. Chako's mission is to foster social, environmental and economic opportunities on the island of Zanzibar by enabling local artisans to preserve their craft, explore their creativity, and live a more sustainable life while reducing the waste on the island. Anneloes is also a part-time lecturer of Tourism Management at Zeeland University of Applied Sciences in The Netherlands.

Valentina Rognoli is Associate Professor in Design at Politecnico di Milano, Italy. She is a pioneer in the field of Materials for Design and established an internationally recognised expertise on the topic in research and education. Her research and teaching activities focus on Materials for Ecological transition; materials; Bio-based and fabricated Materials; Materials from Waste; Materials for interactions and IoT (ICS Materials); Speculative Materials; Tinkering with materials, Materials Driven Design method, CMF design, emerging Materials Experiences, and material education in the field of Design. She is one of the founders of Materials Experience Lab (2015).

Polokano Sekonopo is Lecturer at the University of Botswana in the Department of Industrial Design, Faculty of Engineering and Technology. Polokano's research interests include product innovation, inclusive design and renewable energy. He keeps his skills honed by tinkering with both electrical and mechanical products. He has over two decades of experience as an educator of product design. Some of the courses he offers include new product development, CAD/CAM, and product analysis. He holds membership in various professional and community organisations.

Joseph Shade is an emerging industrial designer and product design student at De Montfort University (DMU) in Leicester, UK. He has industry experience in furniture and product design, 3D Printing, and consumer electronics. He has a strong interest in sustainability and design for repair, which he has explored through an Enterprise Placement Year.

Pammi Sinha is Associate Professor in Fashion Management within the School of Design at the University of Leeds, UK. She has undertaken research funded by the AHRC, the Gatsby Trust (Dar es Salaam), Oakdene Hollins, the ESRC and Research England. Having trained as a fashion designer, her research interests are in the links between design and the market, design management strategies, remanufactured fashion design process, postconsumer textile waste management and sustainability of heritage crafts. Recent projects include (i) engaging the Leeds City Region public in understanding textile waste management, and (ii) examining post-consumer textile waste in Colombo, Sri Lanka.

Lucia Sithole is Lecturer in the School of Art and Design at Chinhoyi University of Technology, Zimbabwe. Her main research interests include fashion marketing, portfolio development, social psychology in dress and circular fashion. She has published several articles in the areas and is working on other projects that promote sustainable fashion practices.

Theodore Sloos is a Master's student in Integrated Product Design at KTH Royal Institute of Technology, Sweden and a graduate of BSc Product Design at De Montfort University (DMU), UK. Theodore completed the DMU Frontrunners Internship on Computer-Aided Design and Upcycling between 2022 and 2023, supervised by Dr Kyungeun Sung.

Pranoti Sonawane is a Senior Product Designer at Cognizant, where she specializes in creating intuitive user experiences for complex B2B products. She holds a Master in Design, and excels at simplifying intricate problems into user-friendly designs. Pranoti is deeply committed to sustainability, ensuring her design process and work is economical, sustainable, and scalable. Her notable achievements include winning the MIT Hackathon and several other national and international competitions with her team. Outside of her professional endeavours, Pranoti enjoys exploring new technologies and contributing to the design community, reflecting her passion for innovation and creativity.

Kyungeun Sung is Associate Professor in Design at De Montfort University, UK. She has research interests and expertise in sustainable design, craft, production, business, supply chains, behaviour, and consumption, focusing on upcycling, circular economy and net zero. Her research uses design as a tool for new product development, service innovation, behaviour change, scaling up niche sustainable practices/behaviours, and transition to the circular economy and net zero. Her recent books are *Research Journeys to Net Zero: Current and Future Leaders* (2024) and *State-of-the-Art Upcycling Research and Practice* (2021). She has published over thirty peer-reviewed academic papers and book chapters.

Nnyaladzi Tema is a Professional Urban Planner and Urban Designer and Lecturer of Urban Planning and design in the Department of Architecture and Planning at the University of Botswana. He is a recipient of the Royal Institute of

Town Planning Book Prize for outstanding performance for his master's degree in Sustainable Place Making and Urban Design at Kingston University-London. He is an emerging researcher with an interest in sustainable places and urban design frameworks for developing countries. His published research work focuses on enabling equitable access to water infrastructure, particularly to low-income communities.

Abhishek Tiwary is Associate Professor of Engineering Management at the School of Engineering and Sustainable Development at De Montfort University, UK. He is the recipient of the Marie Curie Fellowship and the Royal Academy of Engineering Industrial Fellowship to pursue interdisciplinary research and methods. His current research involves system-scale sustainability assessment of technology innovations, both in the UK and in international settings. He has been involved in numerous funded projects evaluating the feasibility of environmentally benign solutions to global challenges on pollution control, resource reuse and renewable energy for both industrial and societal applications.

Thomas Wylam is a Product Design Engineer at SharkNinja, UK, and a graduate of BA Product Design at De Montfort University (DMU), UK. In 2023, Thomas worked on the NPD4CE (New Product Development for Circular Economy and Upcycling) project, which was funded by DMU HEIF (Higher Education Innovation Funding) and supervised by Dr Kyungeun Sung, Mik Pieniazek, and Dr Abhishek Tiwary.

Yi Zhong is a PhD candidate in the Faculty of Arts, Design and Humanities at De Montfort University in Leicester, UK. Her research interests are in intercultural communication, service design, sustainable design, cultural design, behavioural change, and health and wellbeing. Her current PhD research topic is focused on improving the international student services in UK higher education institutes to improve international students' intercultural adaptation experience. She has experience in working as a Research Associate in Circular Economy at De Montfort University and currently works as a Research Associate for the 'Upcycling for everyone' project.

PREFACE

Upcycling is a design, business and social practice that deals with waste and transforms it into new products or materials of higher quality and/or value than the compositional elements. Upcycling reduces solid wastes, increases resource/material efficiency, and lowers industrial energy consumption involved in mass production with virgin materials. Therefore, it contributes to the reduction of greenhouse gas emissions, thus transitioning towards net zero. It creates new job opportunities for businesses and contributes to people's financial and emotional/psychological well-being in households. Despite the benefits, upcycling remains a niche practice, calling for scaling up – making it a mainstream or business-as-usual practice – in order to realise the full potential of upcycling. Whereas the recent increase in publications, initiatives, and upcycling businesses reflects growing academic and industrial interest, research is still in its infancy. Development of upcycling theory and practice across industries, disciplines and countries is required for growth.

This book presents the key findings of the International Upcycling Research Network (IURN) at the forefront of global upcycling research, practice, and opinion. The aim of the IURN was to expand our understanding of current upcycling research and practice across industries, disciplines, and countries and to develop new cross-industry, multidisciplinary, international collaborative research projects and initiatives. IURN is funded by UKRI (UK Research and Innovation) and AHRC (Arts and Humanities Research Council) between 2022 and 2024. The network included over 70 global experts (both academics and practitioners) in upcycling and related fields such as circular economy, renewable energy and sustainable design, engineering and businesses. These experts, with their wealth of knowledge and experience, have contributed chapters that present diverse theoretical and empirical upcycling research and practices in fashion and textiles, products, the built environment, and the community.

The book offers new theories, perspectives, approaches, and scientific research findings related to upcycling. It presents the best upcycling practices across industries and countries with ample images, illustrations, diagrams, graphs, and tables. The contributors propose a sustainable alternative to conventional manufacturing and consumption patterns, contributing to a more circular economy and a greener future. Importantly, these contributions come from six continents: Africa, Asia, Australia, Europe and North and South America and fifteen countries: Australia, Botswana, Brazil, Canada, Colombia, Hong Kong, India, Italy, Kenya, South Korea, Sweden, Tanzania, UK, USA, and Zimbabwe, reflecting a wide variety of current upcycling research, practices, and opinions in different geographical, economic, socio-cultural, and technological contexts at the international level. The main benefit of reading this book is to be informed by the latest upcycling research, practice, ideas and concepts on upcycling by global experts, fostering a sense of global community and shared purpose.

FOREWORD

In an era marked by urgent environmental challenges and the pressing need for sustainable solutions, 'Upcycling Research: Global Perspectives on Product, Fashion and Theory' emerges as a timely and crucial contribution to the field of sustainability and circular economy.

This comprehensive volume brings together a rich tapestry of research, case studies and insights from across the globe, offering a multifaceted exploration of upcycling practices and their potential to reshape our approach to design, production and consumption. What sets this book apart is not only its global perspective but also the diverse backgrounds of its contributors. Hailing from academia, industry and non-profit organizations, these experts offer unique insights into the practical applications of upcycling, backed by rigorous research and real-world case studies.

The book spans a diverse range of industries and applications, demonstrating the versatility and far-reaching impact of upcycling principles. In the realm of fashion, we explore how designers are turning waste into wearable art, challenging the 'fast fashion' paradigm and promoting a more circular approach to clothing production. The product design sections delve into innovative ways of creating functional and aesthetically pleasing items from materials that would otherwise end up in landfill. In examining the built environment, the book showcases how architects, engineers and urban planners are integrating upcycled materials into sustainable building practices, contributing to more resilient and eco-friendly cities.

By featuring studies and initiatives from countries as diverse as Kenya, Colombia, Hong Kong, Thailand, Brazil, Zimbabwe and Botswana, alongside examples from the UK and other developed nations, it provides a nuanced understanding of how upcycling can be adapted and implemented in various cultural and economic contexts. This global outlook is particularly valuable as we seek solutions that can be scaled and replicated across different regions to address our shared environmental challenges.

The book doesn't shy away from the complexities and challenges of implementing upcycling practices. It delves into the technical, economic, and social aspects of upcycling, offering critical insights into both the opportunities and obstacles faced by practitioners, businesses, and educators. By examining these challenges head on, the book provides a realistic and practical guide for those looking to integrate upcycling into their work or studies.

Perhaps most importantly, this book goes beyond the material aspects of upcycling to explore its potential for social impact. The concept of 'social upcycling' – applying upcycling principles to empower marginalised individuals and communities – is a particularly innovative and thought-provoking contribution that expands our understanding of what upcycling can achieve.

As educators, researchers, designers, and policymakers grapple with the urgent need to transition to more sustainable and circular economic models, this book serves as an invaluable resource. It not only provides a comprehensive overview of current upcycling practices and research but also points the way forward, identifying areas for future investigation and development.

Most importantly this book is more than just a collection of studies – it's a call to action. It challenges us to rethink our relationship with materials, waste and design and offers practical pathways for creating a more sustainable and equitable future. Whether you're a student, practitioner, researcher, or simply someone interested in sustainable solutions, this book offers insights and inspiration for re-imagining our world through the lens of upcycling.

As you explore these pages, prepare to be challenged, inspired and equipped with the knowledge to contribute to this vital field. The future of our planet depends on our ability to innovate and implement sustainable practices – and upcycling, as this book compellingly argues – has a crucial role to play in that future.

I hope that this book inspires and informs researchers, practitioners, and anyone interested in sustainable development. The diverse perspectives and case studies presented here demonstrate that upcycling is more than just a trend – it is a vital strategy for addressing some of the most pressing issues of our time. As we face the challenges of the twenty-first century, let us embrace the principles of upcycling and discover the rich tapestry of ideas, research, and creativity that contribute to shaping and building a more sustainable and responsible future.

Professor Tracy Bhamra
Provost and Pro-Vice-Chancellor (Global)
Royal Holloway, University of London

ACKNOWLEDGEMENTS

We would like to thank all the authors who have contributed to and presented at the International Upcycling Festival 2024 at De Montfort University, UK. This project has been funded by the UKRI (UK Research and Innovation) AHRC (Arts and Humanities Research Council) Research Networking Grant for the International Upcycling Research Network (AH/W007134/1). We are grateful for the Research Award at De Montfort University, which has allowed us the time to devote to this book. We would also like to express our sincerest gratitude to our families and friends for their unwavering love and support throughout the process.

ABBREVIATIONS

3D	3-dimensional
ABR	Arts-Based Research
AHRC	Arts and Humanities Research Council
ANOVA	Analysis of variance
ASEAN	Association of Southeast Asian Nations
B2B	Business to Business
BREEAM	Building Research Establishment Environmental Assessment Method
c-Si	crystalline silicon
CBM	Community Based Solid Waste Management
CE	Circular Economy
CFRP	Carbon Fibre Reinforced Polymer
CNC	computer numerical control
CO_2	carbon dioxide
CSR	Corporate Social Responsibility
DEA	Department of Environmental Affairs
DEFRA	Department for Environment, Food & Rural Affairs
DEQP	Department of Environmental Quality Promotion
DIY	Do It Yourself
DMU	De Montfort University
EEE	electro electronic equipment
EU	European Union
EVs	Electric Vehicles
EVA	ethylene vinyl acetate
FAO	Food and Agriculture Organization
FLW	Food Loss and Waste
FSC	Food Supply Chain
GHG	greenhouse gases
HEIF	Higher Education Innovation Funding
IKS	indigenous knowledge systems
IP	Internet Protocol
ISM	Interpretive Structural Modelling
IUF	International Upcycling Festival
IURN	International Upcycling Research Network
LCB	Leicester Creative Business
LED	light-emitting diode
LEED	Leadership in Energy and Environmental Design
MICMAC	Matriced' Impacts Croise's Multiplication Applique and Classment
MSW	Municipal Solid Waste
NGO	Non-Governmental Organisation
NPD4CE	New Product Development for Upcycling and Circular Economy

OECD	Organisation for Economic Co-operation and Development
OW	Organic Waste
PESTEL	Political, Economic, Social, Technological, Environmental, and Legal
PV	photovoltaics
R&D	Research and Development
RM	Reachability Matrix
RRR	Reused, Refurbished or Remanufactured
RTW	Ready-to-wear
SDGs	Sustainable Development Goals
SMEs	Small and Medium Enterprises
SPSS	Statistical Package for the Social Sciences
SSIM	Structural Self-Interaction Matrix
SWOT	Strengths, Weaknesses, Opportunities and Threats
UKRI	UK Research and Innovation
UN	United Nations
UNHCR	United Nations High Commissioner for Refugees
UVARs	Urban Vehicle Access Regulations
VAT	value-added tax
WEEE	Waste Electrical and Electronic Equipment

Introduction

UPCYCLING RESEARCH WITH GLOBAL PERSPECTIVES ON FASHION, PRODUCTS, THE BUILT ENVIRONMENT, COMMUNITY AND EDUCATION

Kyungeun Sung and Richie Moalosi

Upcycling is a design, business and social practice that deals with waste to transform it into new products, artefacts or materials of higher quality and/or value than its compositional elements (Sung, 2015; 2017). It is an umbrella concept that encompasses a number of core circular economy practices (Stahel, 2016), such as creative and innovative forms of reuse, repair, refurbishment, remanufacturing, redesign and remake. Upcycling reduces solid wastes, increases resource or material efficiency and lowers industrial energy consumption involved in mass production with virgin materials. It contributes to the reduction of greenhouse gas emissions, thus transitioning towards net zero (Sung, 2024; 2017). It creates new job opportunities for businesses, especially involving disadvantaged groups of people. It contributes to people's financial well-being by helping them to save and make money, as well as to their emotional/psychological well-being, creativity and sense of achievement in households (Singh et al., 2019; Sung, Cooper & Kettley, 2014). Despite the benefits, upcycling remains largely a niche practice. Scaling up upcycling (i.e., making it a mainstream or business-as-usual practice) (Hartmann & Linn, 2008) could realize its full potential. Whereas the recent increase in publications, initiatives and upcycling businesses reflects growing academic and industrial interest, research is still in its infancy. Development of upcycling theory and practice across industries, disciplines and countries is required for growth (Sung, Singh & Bridgens, 2021).

Recognizing the knowledge gap, the International Upcycling Research Network (IURN), funded by UKRI (UK Research and Innovation) AHRC (Arts and Humanities Research Council), aimed to increase our understanding of current upcycling research and practice across industries, disciplines, and countries and to develop new cross-industry, multidisciplinary, international collaborative research projects and initiatives. The IURN project was led by Dr Kyungeun Sung (Principal Investigator) at De Montfort University (DMU) and Prof Richie Moalosi (Co-Investigator) at the

University of Botswana (UB) between 2022 and 2024 (Moalosi & Sung, 2022). The network included over 70 global experts (both academics and practitioners) in upcycling and related fields such as circular economy, renewable energy and sustainable design, engineering and businesses. The culmination of the IURN project was the International Upcycling Festival (IUF) 2024, where some of the world's experts contributed academic papers presenting diverse theoretical and empirical upcycling research and practice in fashion and textiles, products, the built environment, community, and education. This book is the IUF2024 Proceedings, which contains the key findings of the IURN at the forefront of global upcycling research, practice, and opinion.

Rationale

In an era defined by environmental catastrophe, resource scarcity and a growing consciousness of sustainability, upcycling has emerged as a transformative practice that bridges creativity, innovation, and ecological responsibility. The book *Upcycling Research: Global Perspectives on Product, Fashion and Theory* explores the dynamic field through a multifaceted lens, offering a comprehensive examination of upcycling as a global phenomenon. This book offers a critical exploration of how upcycling is reshaping industries, cultures, and theoretical frameworks globally.

The rationale for this book is anchored on three key pillars:

I. *The imperativeness of sustainable practices*: As the globe grapples with the ramifications of overconsumption and waste, upcycling presents a promising solution to reduce environmental impact. This book highlights upcycling prospects to disrupt traditional linear economies and foster circular systems. It aligns with the principles of the circular economy, as articulated by the Ellen MacArthur Foundation (2015), emphasizing the need to design waste out of systems and regenerate natural ecosystems. By presenting research and case studies from diverse continents, the book underscores the global relevance of upcycling in addressing sustainability challenges.

II. *Cultural and theoretical facets*: Upcycling offers practical solutions to challenges, and it is a cultural and theoretical movement that challenges traditional notions of value, waste and creativity. The book draws on frameworks developed by McDonough and Braungart's (2002) 'cradle-to-cradle' design and Fletcher's (2014) 'slow fashion'. It explores the philosophical and theoretical underpinnings of upcycling, examining how it intersects with concepts such as post-consumerism, the use of less of the world's dwindling resources, and the appreciation of the world's natural and built environments. It also investigates how upcycling practices vary across regional cultures, reflecting local traditions, materials, and aesthetics, thereby offering a rich tapestry of global perspectives. By weaving together theoretical insights and cultural narratives, this book offers a rich, interdisciplinary understanding of upcycling as a global phenomenon.

III. *Innovation and future directions:* Upcycling provides a rich environment for innovation, blending traditional craftsmanship with cutting-edge technology. Nonetheless, challenges remain in scaling upcycling practices and integrating them into mainstream supply chains. This book addresses these challenges by showcasing pioneering research and emerging trends from small-scale artisanal practices and scalable businesses that can be replicated. This visionary book aims to inspire designers, researchers, and policymakers to push the boundaries of what upcycling can achieve.

Upcycling Research: Global Perspectives on Product, Fashion and Theory is a timely and vital contribution to the fields of sustainability, engineering, design, business and cultural studies. It offers a multidisciplinary view of upcycling, weaving together practical applications, theoretical insights, and global narratives to provide a deeper understanding of its importance in the twenty-first century. This book is a call to reimagine our relationship with materials, waste, innovation and creativity, urging readers to see the potential in what is often discarded and to envision a more sustainable and equitable future.

Five themes

Five themes emerged from the contributions, which form the five parts of this book: (i) Fashion and textiles (ii) Engineering and the built environment (iii) Design and business (iv) People and community and (v) Design education. Although the main benefit of upcycling is considered to be environmental, some chapters in this book show social impact is often the main benefit (e.g., Chapters 4 and 15).

Part I, Fashion and Textiles, is opened by Yufei Hu, Pammi Sinha and Eunsuk Hur who looked at emotional design for fashion upcycling – more specifically, how design research could help fashion upcycling practitioners meet consumers' emotional need for upcycled fashion. They conducted a systematic literature review and identified five factors that contribute to a consumer's emotional experience: aesthetics, wearability, the memories evoked in the wearer, the wearer's positive perception of look and feel and cultural elements. In Chapter 2, Sophia N. Njeru and Richie Moalosi introduce Kenyan fashion designers' knowledge, attitude and practice of upcycling. They studied eight Kenyan fashion SMEs and identified valuable lessons for designing interventions that could support fashion designers in a paradigm shift towards upcycling for socio-cultural, economic and environmental benefits. Chapter 3 by Chanjuan Chen and Trina Harlow demonstrates the use of stitching and upcycling to transform the experiences of refugee children into empowerment. Their art-based research showed that upcycling could not only use waste materials for environmental benefit but can also be used as a medium for powerful storytelling and calls to action and empowerment, focusing particularly on marginalized populations (in this case, refugee camp children). In Chapter 4, Edlight Mutungwe, Walter Chipambwa and Lucia Sithole explored upcycling fashion practices in Harare (the capital of Zimbabwe) to understand the critical motivations for fashion upcycling in a developing economic context. They studied fashion experts from bespoke fashion retailers and the garment manufacturing industry and found that there is a need to raise consumer awareness and to improve the skills of fashion designers (e.g., adopting technologies to support the development and marketing of upcycled products). In the final chapter of this part (Chapter 5), Pranoti Sonawane, Lovish Bandwal and Yagnesh Gohil present FABIO, an innovative service that engages stakeholders and fosters a collaborative community, envisioning a future where jeans are treasured possessions that are continuously regenerated through creative upcycling. They show the key results, such as encouraging customer participation in the upcycling process, providing tailored support and DIY options, and incorporating Agraloop technology (i.e., refining natural fibres derived from crops into textile-grade fibres called Agraloop BioFibre) to transform used jeans into biodegradable fabric.

In Part II, Engineering and the Built Environment, Abhishek Tiwary and Dazel Empe provide an engineering management perspective on upcycling carbon-fibre reinforced polymer (CFRP) matrix composites in Chapter 6. They carried out a PESTEL-SWOT analysis (based on political, economic, social, technological, environmental, and legal elements, as well as strengths, weaknesses, opportunities and threats) to identify the key underlying factors. They recommended balancing the role of recycling and upcycling techniques through a clear product development system, with due foresight to consider end-of-life at the design and manufacturing stage, as well as creating a market demand for the upcycled products. In the following chapter (Chapter 7), Kyungeun Sung, Yi Zhong and Patrick Isherwood present the literature review on circular economy solutions for solar photovoltaic (PV) panels and make recommendations for future research for UK academics. Future research should prioritize understanding the current status quo of the PV value chain in the UK, benchmarking the best practices for potential replication in the UK, and developing ready-to-deploy pathways with reproducible processes and supply chain systems with a usable classification and recertification system for PV module waste. In Chapter 8, Nnyaladzi Tema and Benjamin Bolaane explore the possible integration of upcycling into urban infrastructure and the built environment. They argue that upcycling has the potential to create resilient, sustainable urban environments that support environmental integrity, social equity and economic viability, particularly by promoting community engagement. In Chapter 9, Claudia García Flórez introduces a case study of sustainable practices of a Colombian coffee farm. She assessed three coffee by-products (pulp, parchment, and ground coffee waste) in Guaduas, Cundinamarca (a region of Colombia that has recently experienced a surge in coffee plantations). She confirmed the potential of coffee waste upcycling as a promising way to effectively address the complexities of waste management while simultaneously extracting valuable resources from residual coffee by-products.

In Part III, Design and Business, Patrick Pearce, Martin Racine, Isabel Ordóñez and Alice Jarry talk about scaling up upcycling in relation to furniture and object upcycling in Chapter 10. They offer a comparative analysis of 20 upcycled objects to contribute to an understanding of scaling up upcycling at an operational level, focusing on operational scalability and business viability. In Chapter 11, Kyungeun Sung, Amal Abuzeinab and Richie Moalosi unravel vital challenges and critical success factors for scaling up upcycling businesses across industries and countries using Interpretive Structural Modelling (ISM). The findings indicate the necessity of engaging the public in upcycling campaigns, events and initiatives with the objective of increasing awareness and addressing negative perceptions of upcycled products. Furthermore, the findings emphasize the importance of facilitating access to used or waste materials and of addressing labour-intensive processes and inadequate facilities or equipment using appropriate technologies, techniques, and innovative production methods. Chapter 12, by Kevin Cheung and Kyungeun Sung, is about upcycling practice in Hong Kong. The chapter presents several exemplary upcycling products created, and projects organized by Kevin, from Boombottle (waste medicine bottles transformed into portable speakers) to community workshops and installation arts and exhibitions. In Chapter 13, Luca Alessandrini, Valentina Rognoli and Karen Marie Hasling propose a new framework for upcycling organic waste for materials design. The framework is expected to serve as a tool for visualizing a wide range of data, including the types of organic waste, their origins and the processes for creating new circular materials.

In Part IV, People and Community, Joon Sang Baek and Kyungeun Sung introduce a rather controversial topic, the upcycling of throw-away people, in Chapter 14. They argue that upcycling can be applied to human resources to promote the inclusion and added value of economically inactive people. Their comparison between upcycling design and empowerment design shows the potential of design thinking to create environmental and social benefits by revaluing underused products and individuals alike. Chapter 15, by Montida Lertnimanoradee, Busayawan Lam and Eujin Pei, explores successful cases of community participation in waste management and upcycling in Thailand. They reported on four case studies and seven lessons learned from the case studies, highlighting the benefits of participating in community-led waste management. In Chapter 16, Hamilton Ortiz presents the best local reuse action practice in São Paulo, Brazil, called Sucata Quântica, which reuses objects and materials and acts as an educational, eco-design and eco-fabrication movement. Hamilton, as the founder of Sucata Quântica, describes the evolution of the initiative since 2018 – how it has been a driving force behind many upcycling interventions in São Paulo and is continuously working to establish and integrate pilot reuse facilities. In Chapter 17, Walter Chipambwa, Edlight Mutungwe and Chipo Chuma describe their action research on community empowerment through upcycling pre-consumer textile waste. They worked with women in the Musoromuchena community (in Zimbabwe) to upcycle used fabric waste and offcuts to design and make items such as bags, blankets, bed covers and mats. Their project demonstrated that community upcycling of pre-consumer textile waste is a great way to reduce waste, create new products, and save and make money for those involved.

In Part V, Design Education, Richie Moalosi, Polokano Sekonopo, Yaone Rapitsenyane and Paulson Letsholo describe their efforts to incorporate upcycling into the industrial design programme at UB in Chapter 18. Their case study revealed the transformative potential of integrating upcycling into the curriculum, fostering a generation of designers capable of envisioning innovative and environmentally responsible solutions. In Chapter 19, Kyungeun Sung, Joseph Shade, Thomas Wylam, Tony Lorance, Mik Pieniazek and Abhishek Tiwary explain their student design and innovation project on new product development for upcycling and circular economy focusing on everyday electronic products at DMU. The documented project processes and outcomes, as well as student feedback, showed that it was a good learning opportunity for the students. The industry professionals involved in the project also expressed how much they valued the opportunity to contribute to the project and be informed by the innovative ideas. Chapter 20, by Kyungeun Sung, Millie Rees, Theodore Sloos, Richie Moalosi, Yaone Rapitsenyane and Anneloes Roelandschap, looks at another set of student projects at DMU, partly in collaboration with UB and Chako Zanzibar (upcycling social enterprise in Tanzania). Two student projects involved upcycling common waste for product design as an extracurricular activity: Millie used textile upcycling to create a range of textile-based products (e.g., cushions, soft toys) for the DMU Made Christmas Fair and organized a community textile upcycling workshop; and Theodore experimented with different materials (e.g., pallet upcycling to create a chair, plastic and glass upcycling to make new products for Chako). In the final chapter, Chapter 21, Thatayaone Mosepedi and Gerald Kennedy present the opportunities and challenges of implementing upcycling in industrial design education and its potential to create a sustainable, green industry in Botswana, based on their mixed-methods study with educators, students and industry professionals. Their findings show that the integration of upcycling into the industrial design programme has the potential to address environmental concerns, stimulate innovation, and promote sustainable practices in Botswana, while challenges such as lack of awareness, inadequate infrastructure, and limited funding need to be addressed for the successful implementation of upcycling into the curriculum.

References

Ellen MacArthur Foundation. (2015). Towards a Circular Economy: Business Rationale for an Accelerated Transition. Available at: https://www.ellenmacarthurfoundation.org/towards-a-circular-economy-business-rationale-for-an-accelerated-transition.

Fletcher, K. (2014). *Sustainable Fashion and Textiles: Design Journeys*. London: Earthscan Publications Ltd.

Hartmann, A. and Linn, J.F. (2008), 'Scaling up: a framework and lessons for development effectiveness from literature and practice', *Wolfensohn Center for Development Working Paper*, Available at: https://papers.ssrn.com/sol3/papers.cfm?abstract_id=1301625

McDonough, W., & Braungart, M. (2002). *Cradle to Cradle: Remaking the Way We Make Things*. North Point Press.

Moalosi, R. and Sung, K. (2022), 'Promoting Upcycling through an International Research Network', *International Online Conference on Reuse, Recycling, Upcycling, Sustainable Waste Management and Circular Economy.*

Singh, J., Sung, K., Cooper, T., West, K. & Mont, O. (2019), 'Challenges and opportunities for scaling up upcycling businesses – The case of textile and wood upcycling businesses in the UK', *Resources, Conservation and Recycling*, 150: 104439.

Stahel, W.R. (2016), 'The circular economy', *Nature*, 531(7595): 435–438.

Sung, K. (2015), 'A review on upcycling: Current body of literature, knowledge gaps and a way forward', *Proceedings of the 17th International Conference on Environmental, Cultural, Economic and Social Sustainability*, Available at: https://dora.dmu.ac.uk/items/fc497fe1-086c-41bb-a07d-f1384170d6b6

Sung, K. (2017), *Sustainable production and consumption by upcycling: Understanding and scaling up niche environmentally significant behaviour*, Nottingham: Nottingham Trent University.

Sung, K. (2024), 'Sustainable production and consumption by upcycling towards net zero', in K. Sung, P. Isherwood, R. Moalosi (eds), *Research Journeys to Net Zero: Current and Future Leaders*, 79–88, Abingdon: Routledge.

Sung, K., Cooper, T. and Kettley, S. (2014), 'Individual Upcycling Practice: Exploring the Possible Determinants of Upcycling Based on a Literature Review', *The Centre for Sustainable Design*, Available at: https://dora.dmu.ac.uk/items/74d23f26-965c-4c6a-b0d6-414670e9abae

Sung, K., Singh, J. and Bridgens, B. (2021), *State-of-the-Art Upcycling Research and Practice: Proceedings of the International Upcycling Symposium 2020*, Cham: Springer.

Part I

Fashion and Textiles

Chapter 1

EMOTIONAL DESIGN FOR FASHION UPCYCLING

Yufei Hu, Pammi Sinha and Eunsuk Hur

Introduction

The current 'take-make-dispose' linear model (Ellen MacArthur Foundation, 2017) of the fashion industry urgently calls for a transition to a more sustainable circular system (Moorhouse & Moorhouse, 2017). Statistics indicate that out of the 100 billion garments produced by the fashion industry, 92 million tons of clothing end up in landfill annually (Ruiz, 2023). Around 87% of discarded textiles go to landfill, although more than 90% of them are reusable and recyclable (Moazzem et al., 2021).

Transforming the linear model into a circular system is challenging due to the fast-fashion business model, low-price and low-quality clothing coupled with cultural overconsumption (Tripa et al., 2022). Changing the model is imperative for the fashion and textile industry (Chen et al., 2022). Fashion upcycling is a common and effective solution for reusing many fashion and textile products to reduce negative environmental impacts from direct disposal and elevate their value through recycling (DeLong et al., 2017). By transforming discarded items into higher-value creations (Marques et al., 2019), fashion designers can contribute to sustainable fashion based on their creativity and increase the market competitiveness of their products (Cuc & Tripa, 2018).

Although multiple stakeholders contribute to the fashion upcycling industry, the user plays a decisive role in determining whether to acquire a new garment, dispose of an existing one or upcycle it (Fletcher, 2012). Products that consider consumers' emotional experiences have been mentioned frequently in recent years as consumers are more concerned than ever about their emotional needs when purchasing a product (Karaca, 2020). However, there is limited academic research considering the user's emotional experience in fashion upcycling because the focus of the current study is to understand consumers' attitudes towards upcycling and upcycling willingness (Gwilt, 2021; Kim et al., 2021), introduce fashion upcycling strategies (Gam et al., 2011), indicate the demand and advantages of fashion upcycling (James & Kent, 2019) and identify the barriers to fashion upcycling (Zhang & Hale, 2022).

Emotional design can help products evoke positive emotional responses from users (Norman, 2004). Chapman (2009) proposed an emotionally durable design framework to provide a pathway for product designers to create positive user-product relationships by fostering emotional attachment to the product. Emotional attachment and emotional durability have been discussed in fashion design, but the main focus is understanding why the wearer formed an emotional attachment to their clothing (Niinimäki & Koskinen, 2011). Although in product design, considering the emotional aspect of user experience could contribute to the market competitiveness of the product by increasing its attractiveness, the pleasure during usage and potentially cherishing the product (Norman, 2004), the approach, applicability and impact of emotional design in fashion upcycling remains unclear in academic literature.

Considering the consumer's emotional experience is not a new concept in fashion design (Fletcher, 2012), many fashion retailers have used sensory design to trigger their consumers' emotional responses to boost their brand image (Kim et al., 2019). However, most academic research about emotional design is conducted in generic product design. There is a lack of research regarding the potential benefits or impacts when emotional design theory and principles are applied to fashion upcycling. This chapter focuses on how the different applications and benefits of emotional design theory can assist fashion upcycling practices with different dimensions based on upcycling characteristics, building a bridge between emotional design and fashion upcycling. The application of emotional design in other disciplines was reviewed to provide insights into emotional design practices. Moreover, the relationships between wearers, clothing and fashion upcycling approaches were reviewed to explain the correspondence between emotional design and fashion upcycling.

Methods

A systematic review of the literature was conducted. The sample of papers was selected from the Web of Science and Scopus databases, as these two databases have been commonly used as primary sources for literature reviews. The exclusion criteria for both Scopus and Web of Science throughout all searches were papers that were not journal articles, not in English, not accessible, or unrelated to fashion or emotion. The first search used 'emotion and design' and 'memory and design' in December 2022. Thirty-seven papers related to design, emotion and design and memory and design were included (Figure 2.1).

In May 2023, a second search was conducted for papers related to emotional design in fashion, resulting in 11 papers (Figure 2.2 left). Another search was conducted in June 2023, using search terms related to 'fashion upcycling and strategies' resulting in 16 papers (Figure 2.2 right).

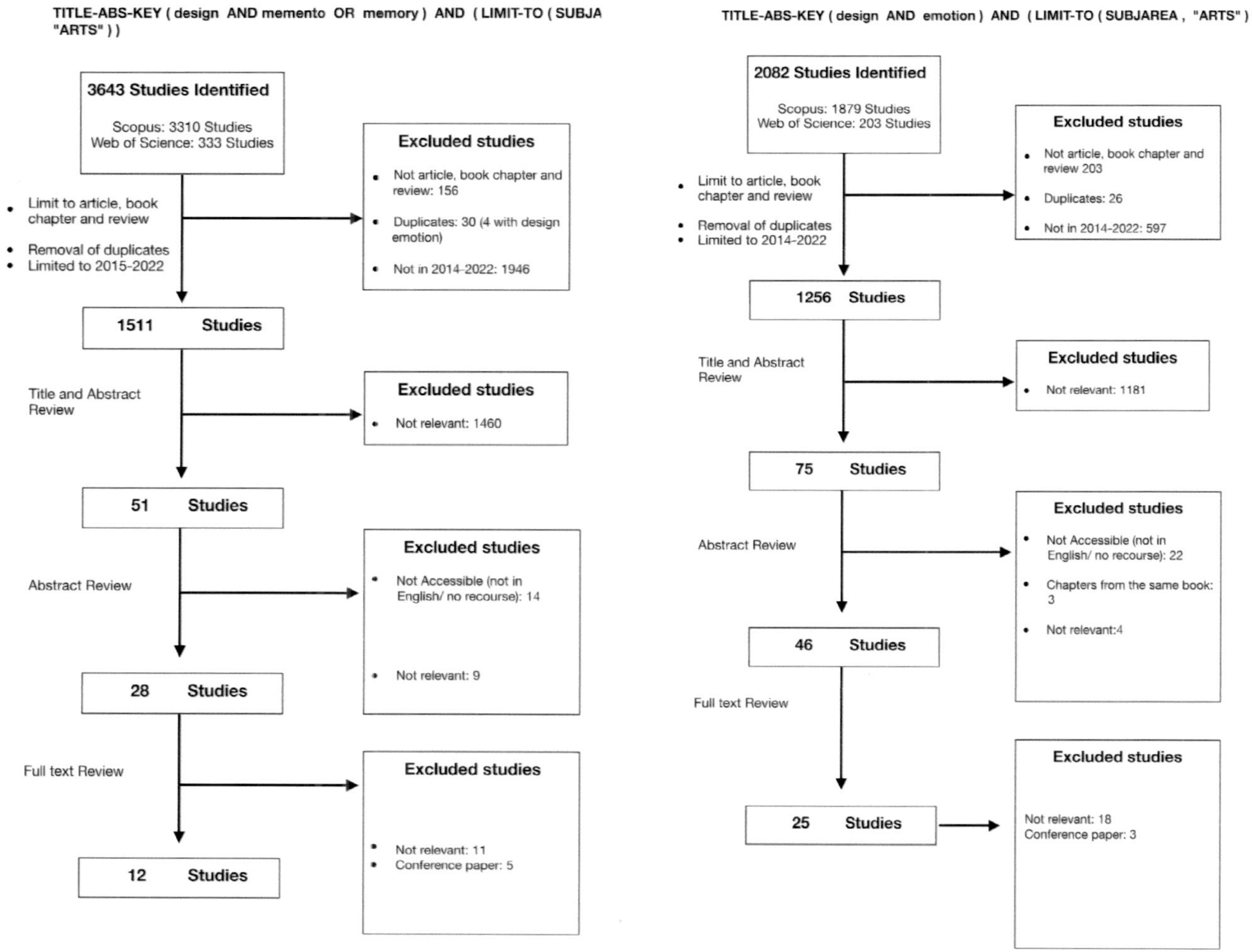

Figure 1.1 Prisma diagram for emotional design.

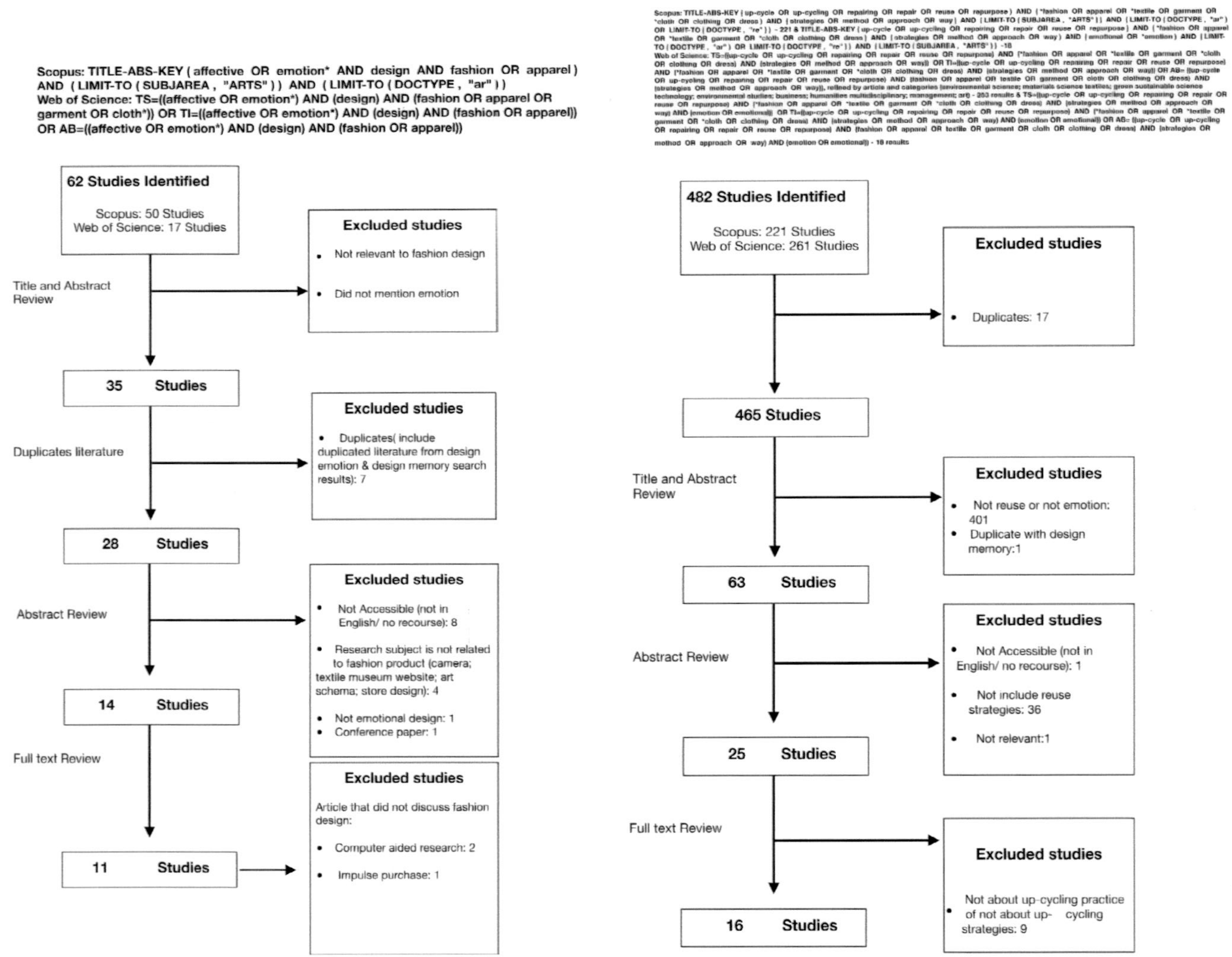

Figure 1.2 Prisma diagram for emotional fashion (left) and Prisma diagram for fashion upcycling (include and exclude emotion) (right).

Results

The searches indicated an uneven distribution in the literature. Most of the papers related to emotional design were conducted in psychiatry, anxiety stress and coping, psychotherapy, audiology and computers in human behaviour. Among the 48 papers identified as relevant to emotional design, most papers (n=37) examined product design, with only 11 on fashion design focusing on understanding wearer-clothing relationships. The search also identified limited results for emotional design in fashion upcycling. Those papers that articulated a theoretical framework for understanding design and emotion tended to adopt the emotional design framework proposed by Norman (2004) to discuss emotional design in fashion upcycling.

Theoretical Framework: Emotional Design Theory

Emotion has the power to influence how individuals perceive and interact with a product and shape an experience (Dewey, 1934). Through emotional design, designers aim to create a deep and meaningful connection between the user and the product by not only meeting the functional needs but also fostering positive and emotional interactions (Norman, 2004). Designers can effectively communicate and express emotions through their creations by carefully considering elements such as shape, colour and texture to evoke positive emotional responses, such as happiness and satisfaction, when using or interacting with the product.

Norman's (2004) emotional design model outlined three levels of how humans process sensory data: visceral, behavioural and reflective. He further indicated that a successful product should elicit positive emotions on these three levels. The visceral level is people's immediate and emotional reaction when they first see the product. At the visceral level, the aesthetic contributes significantly to consumers' emotional reactions, especially for fashion products (Bloch, 1995), as the product's physical features shape the initial impressions towards and attractiveness of a product (Norman, 2004). The next section discusses emotional fashion upcycling based on this model.

Emotional Design in the Context of Fashion Upcycling

Fashion Upcycling Strategies Understanding the complexity of upcycling in fashion design (such as fashion's ever-changing trends) and knowledge about consumers' physical and emotional demands regarding upcycled clothing is crucial when reusing materials and minimizing waste effectively. Therefore, practical and user-friendly methods (Kam & Yoo, 2022) for upcycling must be proposed. Common approaches to facilitate this process are such as creative zero-waste pattern cutting, modularization redesign, redesigning, and re-styling.

Creative zero-waste pattern cutting has been used to minimize waste during the pattern-cutting process (Lewis et al., 2017). In the context of fashion upcycling, a zero-waste pattern effectively reuses the patterns from gathered clothing by using computer software such as CAD to optimize the yield of clothing components (Lewis et al., 2017; Cuc & Tripa, 2018; Tripa et al., 2022). Modularization redesign refers to clothing redesign using parts extracted from old garments with the help of digital technology, such as a group genetic algorithm (Paras et al., 2018). Bigolin et al. (2022) approached modularization redesign more artistically and poetically by using cameras to extract and categorize discarded clothing based on their style. Redesigning clothing into other products is also a creative way to upcycle unwanted items. Juciene et al. (2022) redesigned discarded denim into decorative elements and sound insulation in finishing panels for interior design. Cao et al. (2022) repurposed textile waste into products such as phone cases, hats and bags. Re-styling refers to the appearance of clothing that can be altered through various upcycling approaches. For example, the colour can be changed by re-dyeing (Sugiura, 2019) and the structure can be modified by deconstructing and then redesigning the garment or adding decorative elements like embroidery or prints (Bigolin et al., 2022; Paras et al., 2023) to help transform the look of the clothing.

Apart from the approaches listed above, the cultural dimension is important to sustainable fashion design (Kozlowski et al., 2019). In fashion, upcycling involving cultural elements, such as quilting from traditional handicrafts or collaboration with local artisans, could increase the emotional durability of the upcycled product (Kam & Yoo, 2022). Besides, preserving and redesigning based on the unique history of recycled products could be a good promotional strategy for upcycling brands (Deschamps et al., 2017), as retaining the stories and traces of the cloth's previous life has been suggested as a critical requirement for them (Paras et al., 2018).

In summary, fashion upcycling strategies, such as creative zero-waste patterns and modularization redesign, could contribute to optimizing the waste materials. However, relevant design skills and aesthetics are needed to redesign discarded clothing into other products and make them attractive to consumers. Among these strategies, incorporating cultural elements and narratives could evoke emotional responses to the product, discussed in more detail in the following section.

Wearer-Clothing Relationship in the Context of Upcycling Neto and Ferreira (2023) suggest that clothing can be seen as non-human relatives of the wearer, serving as an extension of the self (Belk, 1988). The shared experience between the wearer and the garment creates a significant emotional bond, often driven by the emotions and memories associated with that experience, even though wearers may not always be aware of their emotional attachment towards their garments. The consumer's decision to discard clothing signifies a failed relationship (Chapman, 2007) between them and the garment (Neto & Ferreira, 2023). In an interview with ten women (aged between 29 and 69 in the UK), Burcikova (2019) found that all of them owned one special dress that was unlikely to be discarded easily. These dresses are special for five reasons: aesthetic, wearability, the memories they evoke in the wearer, making the wearer look and feel good, and cultural elements that contribute to the uniqueness of the clothing, which in turn helps to foster a long-term relationship between the consumer and the garment.

Table 2.1 presents the literature review analysis by identifying factors that influence the aesthetics of clothing. These factors include beauty, style, colour, fit, and tactility (Niinimäki & Koskinen, 2011), as well as the design and placement of clothing components and elements such as silhouette, length, and waist (Qin et al., 2017; Chen et al., 2020a).

The comfort of the clothing material, function, and quality (durability) of the clothing significantly influence the wearer's wearing experience (Niinimäki & Koskinen, 2011; Armstrong et al., 2016), providing a foundation for time long-term use, contributing to consumer satisfaction towards the clothing. An & Lee (2006) further explored Asian women's emotional needs and found that they would pay more attention to the softness and weight of the material. The clothing makes them look and feel good: a sense of satisfaction and delight in the clothes makes them look pretty and attractive (Armstrong et al., 2016), which usually comes from the design or fitness of the clothing and wearing occasions.

Cultural significance has been considered an essential element because some clothing represents cultural symbolism (Liu et al., 2022), such as the embroidery on the Chinese bellyband representing love and family bonds (Hewei, 2023). Clothing evokes many memories, and it 'links to past, present and future' (Niinimäki & Koskinen, 2011, p. 17). It has been argued that this attachment will more likely appear on essential and meaningful objects; therefore, they are unlikely to be discarded. Emotionally meaningful clothing may not be in active use anymore, but the wearer keeps them because of the memories associated with special or meaningful experiences (Niinimäki & Koskinen, 2011; Armstrong et al., 2016). Table 2.2 presents an overview of the analysis of the literature reviewed for emotional values attributed to memories.

Table 1.1 Factors influencing the aesthetic of clothing

Factors	Author and year	Publisher
Beauty, style, colour, fit and tactility (comfortable material)	(Niinimäki & Koskinen, 2011)	*The Design Journal*
Silhouette, clothing length, top fly, collar, waist, hem, pocket, pleat	(Chen et al., 2020a)	*International Journal of Clothing*
Shoulder width, shoulder slope, waist ease, waist length	(Qin et al., 2017)	*Journal of Interdisciplinary Mathematics*

Table 1.2 Emotional values attributed to memories

Emotional values attributed to memories	Author and year	Publisher
History/past, place, people, moments, childhood; Family ties; Positive associations such as soft tactile feeling	(Niinimäki & Koskinen, 2011)	*The Design Journal*
Special occasion, vacation, youth, childhood, a reminder of home, unexpected gift, a reminder of pleasure, a reminder of love obtained for a special occasion, a reminder of youth, the first piece, a reminder of a special person received from a special person; Reference group (family ties, represent a specific thing or people); Expression of self (expression of self, expression of one's ideology, positive association, made for me)	(Armstrong et al., 2016)	*The Design Journal*

Theorizing Emotional Fashion Upcycling Design

Norman stated that all three levels of design – visceral, behavioural and reflective – are closely interlinked (Norman, 2004). As discussed previously, research related to emotional fashion identified five factors that make clothing special: (aesthetic, wearability, makes the wearer look and feel good, cultural elements and the memories they evoke). Fashion upcycling practitioners should consider different strategies to evoke consumers' emotional responses based on these five factors to increase their satisfaction towards upcycled clothing or products.

In fashion upcycling, the appearance, including silhouette, colour, structure, material, print and decoration of clothing, contributes mainly to the consumer's decision, perception and preference on the visceral level (Niinimäki & Koskinen, 2011; Qin et al., 2017; Sugiura, 2019; Chen et al., 2020b) as the visual attractiveness is a crucial aspect in fashion design (Bloch, 1995; Haug, 2016). Therefore, the visceral level of emotional design can be seen as the appearance of fashion-upcycled products. Fashion upcycling strategies that focus on changing the appearance of clothing, such as redesigning the silhouette, redyeing, and reprinting, could contribute to the consumer's emotional experience and satisfaction with the upcycled product.

Behaviour design aims to create positive emotional experiences to enhance user satisfaction and engagement through the usability and functionality of the product (Norman, 2004); it involves a product's function, understandability, usability, and physical feel. A good behavioural design should be human centred as the designer should consider the actual needs of consumers. It should be user-friendly (easy to use) and can fulfil needs. On the behaviour level, consumers consider clothing quality, fit and comfort as significant factors influencing their wearing experience, and researchers have emphasised their significance as they determine if users decide to keep or discard the item based on the interviews conducted in European countries (Niinimäki & Koskinen, 2011; Armstrong et al., 2016). Similarly, research based on aesthetic, emotional, expressive and functional emotional needs shows that Asian women tend to evaluate clothing comfort based on factors such as material softness and weight as the standard for fulfilling their functional and emotional needs (An & Lee, 2015). In fashion upcycling, the durability and quality of the clothing influences consumers' attitudes towards it (Niinimäki & Koskinen, 2011; Armstrong et al., 2016). Therefore, the behaviour level of emotional design can be seen as the wearability of fashion-upcycled products. Fashion upcycling practitioners should consider the actual needs of the user and improve the design based on their feedback. At this level, a successful fashion upcycling design should fit the consumer physically and emotionally. In addition to fitness, the design must make the consumer feel confident when wearing it.

The reflective level is related to the meaning of a product the message it conveys, and the personal memory evoked by it. This level involves individuals' deep thinking, evaluation, and reflection on a product, going beyond its appearance and functionality. At this level, factors such as symbolic meaning, cultural significance, personal identity, and memory associated with the product come into consideration. Factors related to how good the consumers look and feel in the clothing also affect their perceptions (Niinimäki & Koskinen, 2011; Armstrong et al., 2016) on the reflective level.

Figure 2.3 illustrates the potential for the use of memory in emotional upcycling fashion design. While memories are subjective and personal, they can potentially be utilised in several design strategies, such as personalized or customized design (Orth et al., 2018), which suggests that users may cherish this clothing more because of its uniqueness. Consumers who DIY old clothing or brands that provide customized services could utilize the memory shared with clothing and translate it into design elements. For bigger-scale upcycling brands, incorporating narratives,

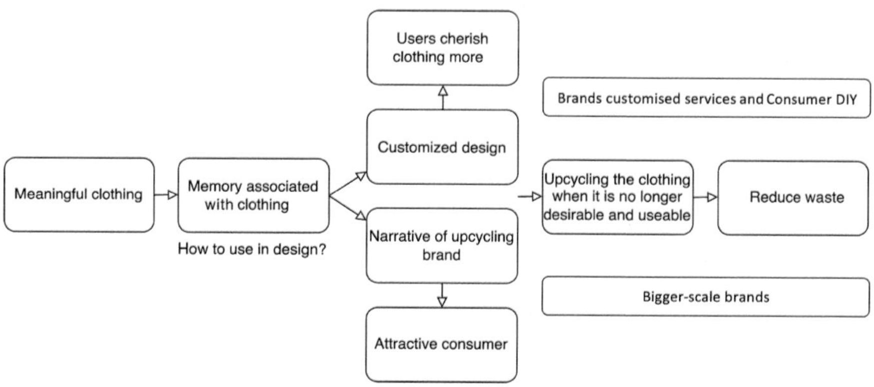

Figure 1.3 Potential use of memory in emotional fashion upcycling design.

such as designing with the patch from previous use, could also contribute to the brand image and competitiveness as consumers may be attracted by the story of the upcycled fashion items (Deschamps et al., 2017).

Building on Norman's (2004) explanation, researchers have identified that other factors, such as culture and memory, also play an important role in fashion upcycling. However, the wearer's memory associated with clothing is personal and formed through long-time usage; therefore, it is hard to design intentionally. Studies have indicated that the more consumers wear a particular garment, the more beautiful and beloved it becomes (Schifferstein & Zwartkruis-Pelgrim, 2008; Neto & Ferreira, 2023) as that frequent and enjoyable use experience creates meaningful memories (Niinimäki & Armstrong, 2013). Researchers have conducted interviews to examine attachment factors for different types of clothing and found that memory played the most significant role in influencing product attachment in dress and suit (Armstrong et al., 2016). At the same time, factors such as function, emotional value, fit and aesthetics were more important for other types of clothing like T-shirts or jeans. Besides memory, cultural elements can also contribute to the consumer's emotional responses and their emotional attachment to clothing (Kam & Yoo, 2022). In addition, cultural elements in upcycled clothing could also evoke a consumer's emotional response, such as nostalgia. However, this needs to be examined within a specific context as different cultural groups respond differently to cultural products. Accordingly, the reflective level of emotional design can be seen as the emotional satisfaction, memory, narrative and cultural elements provided by the fashion-upcycled product.

Conclusion

This chapter provides insights into the relationship between wearer clothing and fashion upcycling. It aligns emotional design elements in fashion design with fashion upcycling strategies. A widely used emotional design model proposed by Norman (2004) has been introduced and considered in terms of the fashion upcycling process. This study identified five key emotional factors in fashion upcycling design: (i) aesthetic appearance (ii) material and wearability (iii) narrative and memories evoked in the wearer (iv) making the wearer's perception of looking and feeling good, or, emotional satisfaction and (v) cultural elements incorporated into the design. The results show that fashion upcycling neglects consumer's emotional experiences. Incorporating emotional design elements into upcycling could potentially significantly improve consumer emotional attachment to fashion garments and increase emotional durability. As this study used peer-reviewed journal articles as references, it may have overlooked emerging innovative practices. Further empirical studies could enhance understanding of the roles of emotional design, thereby enabling the optimisation of product use and longevity, which could ultimately support alternative ways to advance circular fashion practices.

References

An, H. and Lee, I. (2015), 'A conceptual framework for Asian women's emotional needs in fashion design', *International Journal of Fashion Design, Technology and Education*, 8(3): 206–213.

Armstrong, C.M., Niinimäki, K. and Lang, C. (2016), 'towards design recipes to curb the clothing carbohydrate binge', *Design Journal*, 19(1): 159–181.

Belk, R.W. (1988), 'Possessions and the extended self', *Journal of Consumer Research*, 15(2): 139–168.

Bigolin, R., Blomgren, E., Lidstroem, A., de Oliveira, S.M. and Thornquist, C. (2022), 'Material inventories and garment ontologies: Advancing upcycling methods in fashion practice', *Sustainability*, 14(5): 2906.

Bloch, P. (1995), 'Seeking the ideal form: Product design and consumer response', *Journal of Marketing*. 59: 16–29.

Burcikova, M. (2019), 'One dress: Shaping fashion futures through utopian thinking', *Fashion Practice*, 11(3): 328–345.

Cao, H.T., Cobb, K., Yatvitskiy, M., Wolfe, M. and Shen, H.Q. (2022), 'Textile and product development from end-of-use cotton apparel: A study to reclaim value from waste', *Sustainability*, 14(14): 8553.

Chapman, J. (2007), 'Subject object relationships and emotionally durable design', in T. Cooper (ed), *Longer Lasting Products*, 61–76, London: Routledge.

Chapman, J. (2009), 'Design for (emotional) durability', *Design Issues*, 25(4): 29–35.

Chen, J.Y., Polston, K., Nicols, E. and Phung, B. (2022), 'A study on textile recycling in college student residence areas', *Journal of the Textile Institute*, 113(9): 1854–1861.

Cuc, S. and Tripa, S. (2018), 'Redesign and upcycling – a solution for the competitiveness of small and medium-sized enterprises in the clothing industry', *Industria Textila*, 69(1): 31–36.

DeLong, M., Alice Casto, M., Min, S. and Goncu-Berk, G. (2017), 'Exploring an up-cycling design process for apparel design education', *Fashion Practice*, 9(1): 48–68.

Deschamps, T., Carnie, B. and Mao, N. (2017), 'Public consciousness and willingness to embrace ethical consumption of textile products in Mexico', *Textiles and Clothing Sustainability*, 2: 1–16.

Dewey, J. (1934), *Art as Experience*, New York: Capricorn Books.

Ellen MacArthur Foundation (2017), 'A new textiles economy: Redesigning fashion's future', Available at: https://www.ellenmacarthurfoundation.org/a-new-textiles-economy.

Fletcher, K. (2012), 'Durability, fashion, sustainability: The processes and practices of use', *Fashion Practice*, 4(2): 221–238.

Gam, H.J., Cao, H.T., Bennett, J., Helmkamp, C. and Farr, C. (2011), 'Application of design for disassembly in men's jacket A study on sustainable apparel design', *International Journal of Clothing Science and Technology*, 23(2–3): 83–94.

Gwilt, A. (2021), 'Caring for clothes: how and why people maintain garments in regular use', *Continuum*, 35(6): 870–882.

Haug, A. (2016), 'A framework for the experience of product aesthetics', *Design Journal*, 19(5): 809–826.

Hewei, T. (2023), 'Chinese folk embroidery bellyband: The intersection of loving symbol, emotional release, and spiritual needs', *Textile: The Journal of Cloth and Culture*, 22(1): 84–101.

James, A.S.J. and Kent, A. (2019), 'Clothing sustainability and upcycling in Ghana', *Fashion Practice*, 11(3): 375–396.

Juciene, M., Dobilaite, V., Albrektas, D. and Bliudzius, R. (2022), 'Investigation and evaluation of the performance of interior finishing panels made from denim textile waste', *Textile Research Journal*, 92(23–24): 4666–4677.

Kam, S. and Yoo, Y. (2022), 'Practice of sustainable fashion design considering customer emotions and personal tastes', *Frontiers in Psychology*, 13: 976471.

Karaca, E. (2020), 'Emotional landscape design through Kansei/affective engineering', in M. Özyavuz (ed), *Theory and practice in sustainable planning and design: Planning, design, applications*, 429–444, Berlin: Peter Lang.

Kim, I., Jung, H.J. and Lee, Y. (2021), 'Consumers' value and risk perceptions of circular fashion: Comparison between secondhand, upcycled, and recycled clothing', *Sustainability*, 13(3): 1208.

Kim, Y.K. and Sullivan, P. (2019), 'Emotional branding speaks to consumers' heart: the case of fashion brands', *Fashion and Textiles*, 6(1): 1–16.

Kozlowski, A., Bardecki, M. and Searcy, C. (2019), 'Tools for sustainable fashion design: An analysis of their fitness for purpose', *Sustainability*, 11(13): 3581.

Leerkes, E.M. and Siepak, K.J. (2006), 'Attachment linked predictors of women's emotional and cognitive responses to infant distress', *Attachment and Human Development*, 8(1): 11–32.

Lewis, T.L., Park, H., Netravali, A.N. and Trejo, H.X. (2017), 'Closing the loop: a scalable zero-waste model for apparel reuse and recycling', *International Journal of Fashion Design, Technology and Education*, 10(3): 353–362.

Liu, H., Wang, Y., Huang, Z. and Romainoor, N.H. (2022), 'A study on the attractive quality attributes of Guochao T-shirt products based on consumer emotional experience', *Journal of Silk*, 59(2): 55–67.

Marques, A., Moreira, B., Cunha, J. and Moreira, S. (2019), 'From waste to fashion – a fashion upcycling contest', *Procedia CIRP*, 84: 1063–1068.

Moazzem, S., Wang, L., Daver, F. and Crossin, E. (2021), 'Environmental impact of discarded apparel landfilling and recycling', *Resources, Conservation and Recycling*, 166: 105338.

Moorhouse, D. and Moorhouse, D. (2017), 'Sustainable Design: Circular Economy in Fashion and Textiles', *Design Journal*, 20(sup1): S1948–S1959.

Neto, A. and Ferreira, J. (2023), 'Lasting bonds: Understanding wearer-clothing relationships through interpersonal love-theory', *Fashion Theory – Journal of Dress Body and Culture*, 27(5): 677–707.

Niinimäki, K. and Armstrong, C. (2013), 'From pleasure in use to preservation of meaningful memories: A closer look at the sustainability of clothing via longevity and attachment', *International Journal of Fashion Design, Technology and Education*, 6(3): 190–199.

Niinimäki, K. and Koskinen, L. (2011), 'I love this dress, if makes me feel beautiful! Empathic knowledge in sustainable design', *Design Journal*, 14(2): 165–186.

Norman, D.A. (2004). *Emotional Design: Why We Love (Or Hate) Everyday Things*, New York: Basic Books.

Orth, D., Thurgood, C. and van den Hoven, E. (2018), 'Designing objects with meaningful associations', *International Journal of Design*, 12(2): 91–104.

Paras, M.K., Pal, R. and Ekwall, D. (2023), 'An exploratory study of Swedish and Romanian organisations to investigate upcycling practice in the clothing industry', *Research Journal of Textile and Apparel*, Vol. ahead-of-print (No. ahead-of-print).

Paras, M.K., Wang, L.C., Chen, Y., Curteza, A., Pal, R. and Ekwall, D. (2018), 'A sustainable application based on grouping genetic algorithm for modularized redesign model in apparel reverse supply chain', *Sustainability*, 10(9): 3013.

Qin, F., Jiang, Y. and Liu, G.L. (2017), 'A behavior experiment study: Information integration in sensory perception of female suit silhouettes', *Journal of Interdisciplinary Mathematics*, 20(1): 243–254.

Ruiz, A. (2023), '17 Most Worrying Textile Waste Statistics & Facts', Available at: https://theroundup.org/textile-waste-statistics/ (accessed 28 June 2023).

Schifferstein, H.N.J. and Zwartkruis-Pelgrim, E. (2008), 'Consumer-product attachment: Measurement and design implications', *International Journal of Design*, 2(3): 1–13.

Sugiura, M. (2019), 'The mass consumption of refashioned clothes: Re-dyed kimono in post-war Japan', *Business History*, 61(1): 106–121.

Tripa, S., Indrie, L., Zlatev, Z. and Tripa, F. (2022), 'Customized clothes – a sustainable solution for textile waste generated by the clothing industry', *Industria Textila*, 73(3): 275–281.

Zhang, L.S. and Hale, J. (2022), 'Extending the Lifetime of Clothing through Repair and Repurpose: An Investigation of Barriers and Enablers in UK Citizens', *Sustainability*, 14(17): 10821.

Chapter 2

KENYAN FASHION DESIGNERS' KNOWLEDGE, ATTITUDE AND PRACTICE OF UPCYCLING

Sophia N. Njeru and Richie Moalosi

Introduction

The current linear consumption cycle is that people extract natural resources from the environment, design and make products, sell them and, after using them for some time, discard them as waste. Upcycling is disrupting this unsustainable cycle as it offers a mix of upgrading and recycling (Wegener, 2016). Upgrading adds value to products, while recycling promotes the reuse of discarded items. Upcycling is taking disposable items and transforming them into products of added value (ibid). Therefore, upcycling offers a promising sustainable consumption alternative by taking disposed products into a new production and value-creation cycle. The products created through upcycling are better than the original, and they tell sustainability stories such as 'production with zero waste', 'small is beautiful' and 'start local, but think global' (Earley, 2011; Wegener, 2016). By reusing waste components, the need for recycling is decreased. The upcycling process saves the use of new materials, water, and energy, thus addressing resource scarcity, promoting sustainable production and consumption and reducing carbon footprint in the circular economy. Designing upcycled products requires a new radical cultural mindset change from the culture of throwing away to a design approach of component modularity that allows products to be deconstructed and reconstructed (Holtström et al., 2019). It demands innovation and a futuristic creativity approach that merges old garbage and new components to create valuable products of added value.

The fashion industry in Africa is mainly driven by Small and Medium Enterprises (SMEs). SMEs contribute significantly to any country's sustainable economic development, innovation and competitiveness. However, the industry's production processes result in environmental degradation, especially from solid waste and unsustainable production processes (Johnson & Plepys, 2021). Fashion enterprises generate large quantities of biodegradable and non-biodegradable waste that end up in landfills, incinerators and bodies of water bodies (ibid). There is no research on Kenya's fashion industry, specifically among SMEs, which addresses upcycling, a potentially creative and economical means to mitigate unsustainable production, consumption and environmental pollution.

The objective was to assess fashion designers' knowledge, attitudes and practice of upcycling. It focused on upcycling and how fashion designers' knowledge and attitudes impact their practice in varied SMEs in the Kenyan fashion industry. Though apparel products negatively impact the environment at every point along the product's lifecycle, this study was limited to the production stage. The significance of this study stems from its ability to advance sustainable practices, influence positive change within Kenya's fashion industry, and support the worldwide movement towards a more socially and environmentally responsible fashion industry.

Methods

The research adopted a case study design for an in-depth description of a small number of cases (Mouton, 2001). The units of analysis were fashion houses/firms domiciled in diverse parts of Kenya. The fashion industry in Kenya is fragmented and thus lacks statistics on its players' population. Hence, the names of 20 fashion houses were obtained from *FashionFix* and style columns in *Sunday Nation* newspapers between January to July 2022 and six designers whom the researchers knew. The fashion houses were contacted and eight willingly participated in the study. The owner/managers were deemed the key decision-makers and had an in-depth understanding of the enterprises' operations. Semi-structured interviews were conducted using online interview schedules shared with the fashion firms' owners/managers. The interview questions focused on the interviewees' demographic characteristics and their knowledge, attitude, and practice of upcycling. A faculty member reviewed the interview questions for validity. The respondents availed images of upcycled fashion products via email. Qualitative data were analysed using thematic analysis (Ali, 2021). Qualitative content analysis was applied to the images of fashion products (Shava et al., 2022). The analysis provided data on the overall style features and specific design details, such as materials, notions, and construction techniques. The data has been presented in narratives and photographs. The fashion design participants were coded as F1–F6, while the leather design participants were L1–L2.

Results

Participants' Demographics

Table 3.1 shows the participants' demographics. Eight firms participated in the study. There were four females and males respectively. Regarding professional qualifications, six undergraduate degrees were in fashion design, one in commerce and one in philosophy. The Master's degree was in fashion design and marketing.

Participant F6, a philosopher, described himself as a Do-it-yourself (DIY) person. One undergraduate degree and diploma holder never specified their discipline. Table 3.1 shows that seven firms employed 1–10 employees and one firm employed between 11–100 people. Seven firms were new start-ups with one to ten years in operation. Only one firm was mature because it had 11 to 20 years in operation.

All the firms had a vision and mission statement. Some of these recognized sustainable fashion: environment, consumer, economy, quality, and culture. For instance, Participant F2 explained that the 'vision is a sustainable fashion brand by unravelling the negative environmental effects of fashion while saving customers' time and money by providing top quality, affordable clothing nationwide'. Participant F2 added that the mission is 'improving the fashion industry with every stitch, offering superior designs, high-quality products with exceptional price-value and superior service to the consumer'. The vision of participant L1 was

Table 2.1 Demographic characteristics of owner/managers and firms

	Demographic characteristics	N	%
Gender	Female	4	50.0
	Male	4	50.0
	Other	0	0.0
Professional qualifications	Undergraduate degree	6	75.0
	Master's degree	1	12.5
	Diploma	1	12.5
Stage/phase of business	New start-up (1–10 years operation)	7	85.5
	Mature (11–20 years operation)	1	12.5
Number of employees	1–10	7	85.5
	11–100	1	12.5
Vision and mission statements	Stated	8	100.0
	None	0	0.0
Products created (multiple responses allowed)	Clothing/Apparel	6	75.0
	Intimate apparel	0	0.0
	Fashion accessories	2	25.0
Products specialization (multiple responses allowed)	Ready-to-wear (RTW)	6	75.0
	Haute couture	2	25.0
	Avante Garde	0	0.0
Target market (multiple responses allowed)	High class	1	12.5
	Middle class	5	62.5
	Low class	1	12.5
	'Green/eco/sustainable'	2	25.0
	Male	2	25.0
	Female	2	25.0
	Youthful	3	37.5
	Middle age	1	12.5
	Elderly	1	12.5
	Persons living with disabilities	1	12.5
	Plus-size	1	12.5

'We aim to represent the African authenticity across the world'. Participant F4's vision was to 'create fashion pieces that respect our beautiful earth and its inhabitants', and the mission was to 'break the stereotype that sustainable fashion is expensive and make these fashion pieces accessible and affordable to all people'. Our vision and mission, respectively, are 'Beautiful and sustainable apparel' and 'create employment and stable income and discover new ways of greener lifestyle' (Participant F6). The vision and mission statements show that the fashion houses were conscious of sustainability issues.

Most of the firms (6) were in clothing/apparel, and two created fashion accessories. The firms specialized in Ready-to-wear (RTW) products (6) and Haute couture (2), as illustrated in Table 3.1. Regarding fashion products, product-orientated PSS (Product-Service-System) models are highly possible with RTW clothing/apparel. Participant F6 was collaborating with Kazuri Beads to diversify into beadwork jewellery.

Fashion products and target market

A total of six participants specialize in clothing/apparel, and two in fashion accessories. In terms of product categories, five participants specialize in ready-to-wear (RTW) and two in haute couture. Only one participant creates both RTW and haute couture. Product-orientated S.PSS (Sustainable Product-Service-System) models are highly possible with RTW clothing/apparel. Participant F6 was collaborating with Kazuri Beads to diversify into beadwork jewellery.

Almost all the participants had one specific market related to either gender or economic class, specifically the middle class. Participant F6 had a combination of markets varying from youth/male/female/green/eco/sustainable and those who yearn for the story behind the apparel. Nonetheless, Participant F2 was focused on high/middle/low class, female/male, green/eco/sustainable, youth, elderly, persons living with disabilities and plus size. The two participants were responsive designers.

Knowledge of Sustainable Fashion and Upcycling

The level of knowledge required to implement upcycling in the fashion industry is high, expressed in their understanding of sustainable fashion and upcycling, its adoption, and its benefits. Most participants' perceptions of sustainable fashion alludes to its environmental, social and economic characteristics such as:

- 'Fashion that does not destroy or affect the environment' (Participant L1).
- 'Consistency, eco-friendly. . .' (Participant L2).
- 'Having eco-friendly and ethical practices and processes of creating fashion products' (Participant F4).
- 'Fashion that is friendly to humanity and environment' (Participant F3).
- 'Clothing that is designed, manufactured, distributed, and used in ways that are environmentally friendly' (Participant F2).

- 'Fashion that is conscious about its effects on the environment it operates in and people who work in it' (Participant F5).
- 'Practice that links clothing from beginning to the end' (Participant F6).

Regarding upcycling, participants stated perspectives that captured its distinct features of pre-and post-consumer waste, value addition, repurposing and producing fashion products. For instance:

- '. . .taking something that is at the end of product life and give it better and new life and value' (Participant F6).
- '. . .the reuse of a product' (Participant L1), 'making use of the leftover product to create something' (Participant F3),
- '. . .using outdated or old products to make new products' (participant FI) and 'recycle to better' (Participant L2).
- '. . .value addition of fashion waste' (Participant F4).
- 'A design-based circular fashion approach, where pre- or post-consumer textile waste material is repurposed or renovated to create new garments' (Participant F2).
- '. . .reusing garments/materials to give them a new life' (Participant F5).

The participants were knowledge-seekers; hence, they obtained information about upcycling from varied sources, including online products, undergraduate fashion design modules on renovated apparel, media, journal articles, faculty, and conversations with industry acquaintances. A unique source for Participant F6, were tailors who considered it to be African culture to upcycle and repair clothes, unlike in Europe, where apparel with a small stain or tear is often discarded. Despite being knowledgeable about upcycling, two participants had not adopted it in creating fashion products because the 'market we serve has not yet taken it up as something they would love' (Participant F5), and it is 'still young in the market and there is limited knowledge on such' (Participant L2). The other six participants have embraced upcycling, recognizing the skills required, the benefits and the tensions involved. The most critical skill required of a designer to master upcycling is creativity, as participant F4 emphasized: 'creating something stylish and wearable from a used piece requires a unique eye and a creative mind'. Other skills required include innovation, imagination, uniqueness, excellent craftsmanship to deconstruct and construct a fashion product, an understanding of the environmental impact of raw materials, knowledge of fabric quality and expertise in the use of work tools.

Attitude toward Upcycling

The participants enjoyed several benefits from upcycling, namely, support for a worthy cause from eco-conscious stylists, designers, and organizations, happiness that their first upcycled collection sold out in two weeks, and high demand for upcycled bags. 'More income is generated from clients who want repairs done but surprise them with a creative design and referral clients who end up ordering a new product' (Participant F2). Participant F6 mentioned, 'creating awareness that people can do upcycling, no mass-production leading to the uniqueness and more people, especially clients, want to champion upcycling'. Despite the benefits, the participants experience a few tensions, such as poor-quality denim products and other second-hand apparel. Some clients are unwilling to spend much money on old apparel mainly because there is little information on sustainable fashion, particularly upcycling. In addition, there is an occasional loss of designers' creations in transit from Africa to Europe, and there is the challenge of quality control, especially for participants who are based abroad. Participant F6 echoed that upcycling only allows them to produce limited editions of their apparel.

It was prudent to establish the opinion of the participants' employees concerning upcycling because they are charged with implementing it in production. The opinions were predominantly positive, such as becoming more interested in such unique products, leading to taking up waste fabric for their use. The employees are fascinated and have a sense of belonging since it is a worthy cause. The employees deem upcycling as a very regular or natural concept as most Africans already practise it, but they were unaware of the terminology (Participant F6). However, a minority of employees consider upcycling to be repairs so are unwilling to give it the attention it deserves.

The Practice of Upcycling

All eight participants produce their creations in-house, thus managing their solid waste. Out of these eight participants, five of them upcycle their solid waste. Notably, none of the participants burned solid waste, indicating their environmental consciousness. The firms were located in urban centres, except for Participant F6, which is domiciled in a rural town and has access to garbage collection services. Those participants that create upcycled apparel mainly use fabric off-cuts of 100% cotton, such as Kente and Kitenge, in addition to polyester, wool and their blends (Figure 3.1).

Second-hand clothes, curtains, and duvets are other raw materials that have been upcycled. Fashion accessory designers utilize leather, some of which is waste from tanneries. Depending on the products created, fashion designers' significant solid waste is fabric off-cuts, in addition to thread cones and paper-and-packaging materials such as cartons. Fashion accessory designers have leather heaps, which Participant L1 states 'if not upcycled could end up in water catchment areas endangering marine life'.

The participants embrace zero solid waste in their operations primarily by using fabric off-cuts to create throw cushions, pillows and pin cushions or sell fabric off-cuts to people who produce face masks, mattresses, and cushions. Participant L1 stressed, 'We upcycle our raw materials, and even the smallest pieces are used to make small accessories such as keyholders'. The positiveness of Participant F6 was 'seeing great potential in solid waste and using it as much as possible'.

Figure 2.1 Tops made from a second-hand bedsheet by Hilda Karanja Wangui.

The process of upcycling from ideation to the final product(s) entails idea generation and concept development using sketching and other means, as well as secondary research on how other designers create garments through upcycling. The process also includes colour rendering, pattern making, sourcing for fabrics and other necessary raw materials such as second-hand curtains, prototype development of two to three sample garments in different colours and sizes, production of final garments, and fitting. Some stages may be skipped, such as prototype development, if a designer is not producing a collection. Other stages are exclusive to Participant F6 in that the finished fashion products are shipped to Berlin, Germany, rather than being sold locally. Fashion products created by upcycling include apparel, namely, two-piece trouser suits that are gender specific or unisex, dresses, blouses, and two-piece skirt suits (Figure 3.2).

Fashion accessories created by upcycling comprise bags, scarves, belts, headbands, shoes, clutch bags, bucket hats and keyholders (Figures 3.3 and 3.4).

The products also include soft furnishings, namely coasters, duvets, table mats and runners (Figure 3.5). Participants created upcycled business-related products, namely patchwork packaging bags, multimedia crafts, pin cushions and rugs used for cleaning equipment.

Discussion

Despite the challenges and considering the perceived benefits of upcycling, more sustainable fashion champions and business models could be encouraged.

Figure 2.2 Two-piece suits made from second-hand curtain fabrics (Aluoch, 2022) (left) and upcycled post-consumer waste (right) by Sidney Nwakanma.

Figure 2.3 Old stilettos renovated with kitenge (left) and a clutch bag and shoes upcycled with kitenge (right) by Caroline Nyarunda.

Figure 2.4 Laptop bag made from a re-sized second-hand bedsheet (left) and an upcycled bedsheet bag (right) by Caroline Nyarunda.

Figure 2.5 Table mats and runners by Caroline Nyarunda.

Kenyan fashion firms, both start-ups and those that are more mature established, irrespective of the owner/manager's gender and professional qualifications, have embedded upcycling into their creations. This is fostered by the owner/managers' in-depth knowledge of and positive attitude towards upcycling. This study's fashion houses were SMEs that offered a wide range of upcycled fashion products in RTW and haute couture to diversified markets. However, the former negates the ethos of sustainability, which frowns on RTW. According to Vezzoli et al. (2022), producing fashion products on order/custom-made ensures firms do not have excess stock/inventory, resulting in waste. The finding concurs with Gurova and Morozova (2018), who state that SMEs significantly contribute to Helsinki's fashion scene and increase its diversity. The designers create a market called the 'sustainable clothing design market,' which combines small-scale designers, production spaces, particular practices, and meanings with sustainable values.

The participants created diverse upcycled fashion products from pre- and post-consumer solid waste in haute couture for the local and international markets. The participants leverage the upcycling opportunities offered to SMEs because they have more flexibility in approaching their products from a lifecycle perspective (Earley et al., 2016). The participants stated that creativity is the predominant skill set required for upcycling, which is embodied uniquely and aesthetically in fashion products and soft furnishings created. This finding echoes Gurova and Morozova (2018), who pre-suppose that designers invest creativity in their products and act as 'entrepreneurs of taste' by adding design ideas for the purpose of creating a new garment from an old one. Two participants acquired their knowledge of upcycling from formal education through modules and faculty. As Gurova and Morozova (2018) assert, one of the biggest problems of sustainable fashion is that designers are often self-taught in the sustainable production of clothes. The designers have to educate themselves, a process that requires time, effort and practice.

The participants of this study can be classified as determiners (creators of boundary objects) and co-creators (facilitators), according to Gardetti (2019). Casto and DeLong (2019) stress that designers should design items that people need rather than those they want to meet both consumer and sustainability needs. A participant's concern that upcycling only allows them to produce limited editions of their apparel should not be deemed as a conundrum. Gurova and Morozova (2018) state that Helsinki's fashion designers are reluctant to use the word 'collection', which is normally aligned with big companies and trends. The designers instead consider it environmentally wise to create timeless, unique products three times a year, with new colours or models from high-quality materials to support the extended life and use of the products.

In the practice of upcycling, the participants make compromises because there is no such thing as being 100 percent sustainable (Gurova & Morozova, 2018). However, it is more important to bring the idea of sustainability to the market, create an alternative niche, educate consumers and develop innovative practices of sustainable production that can bring about future changes with long-lasting positive effects.

Conclusion

The participants were utilizing upcycling to compete in the fashion industry by creating alternatives to unsustainable fast fashion and as a basis for their identity. The participants also contributed creatively and uniquely to humanity's socio-cultural, economic, and environmental well-being, attaining SDGs 9, 12, 13, and 15 and transitioning to the circular economy. The chapter emphasizes the importance of education and collaboration in promoting upcycling practices in Kenya's fashion industry. Designers would benefit from specialized training in upcycling techniques. Collaboration between designers, policymakers and industry stakeholders is crucial to create a supportive ecosystem for sustainable fashion. The chapter has highlighted the potential for upcycling to contribute positively to Kenya's fashion industry in terms of sustainability, cultural expression, and economic growth. However, there are significant challenges that need to be addressed, including limited customer knowledge, awareness and access to materials. By overcoming these obstacles through education, collaboration and policy support, upcycling can emerge as a viable and impactful solution to foster a more sustainable and socially responsible fashion sector in Kenya.

Acknowledgements

The authors acknowledge the participation of the following seven design firms in the study: Ostrich Design Hub, Aytso Bej Fashion House, 2NU Brands, Stylish with Adira, Afrivazi, Rembo International, and Emeka Suits.

References

Ali, M. (2021), 'A systematic literature review of sustainable entrepreneurship with thematic analysis.' *World Journal of Entrepreneurship, Management and Sustainable Development*, 17(4), 742–764.

Aluoch, C. (2022), 'Emeka suits: Addressing pollution'. *Sunday Nation, Lifestyle.*

Casto, M.A. and DeLong, M. (2019), 'Exploring aesthetic response to classic as a means to slow fashion'. *Fashion Practice*, 11(1): 105–131.

Earley, R. (2011), 'Upcycling textiles: Adding value through design', *KEA's Towards Sustainability in the Fashion and Textiles Industry*, 26–27 April, Copenhagen, Available at: http://ualresearchonline.arts.ac.uk/4023/ (accessed 15 May 2023).

Earley, R., Vuletich, C., Goldsworthy, K., Politowicz, K. and Ribul, M. (2016), 'The Textile Toolbox: New Design Thinking, Materials and Processes for Sustainable Fashion Textiles', University of the Arts London, Available at: https://ualresearchonline.arts.ac.uk/id/eprint/11281/ (accessed 16 May 2023).

Johnson, E. and Plepys, A. (2021), 'Product-Service Systems and Sustainability: Analysing the Environmental Impacts of Rental Clothing', *Sustainability* 13: 2118.

Gardetti, M.A. (2019), 'Independent fashion design in Argentina. Is 'Sustainablewashing' a concern'?' in A. Gwilt, A. Payne, E.A. R⊠thschilling (eds), *Global perspectives on sustainable fashion*, 28–40. London: Bloomsbury Visual Arts.

Gurova, O. and Morozova, D. (2018), 'A critical approach to sustainable fashion: Practices of clothing designers in the Kallio neighbourhood of Helsinki'. *Journal of Consumer Culture*, 18(3): 397–413.

Holtström, J., Bjellerup, C. and Eriksson, J. (2019), 'Business model development for sustainable apparel consumption: The case of Houdini Sportswear', *Journal of Strategy and Management*, 12(4): 481–504.

Mouton, J. (2001), *How to succeed in your master's and doctoral studies: A South African guide and resource book*, Pretoria: Van Schaik Publisher.

Shava, G.N., Hleza, S., Tlou, F.N. and Mathonsi, E. (2022), 'Qualitative content analysis', Available at: https://www.researchgate.net/publication/355186712_Qualitative_content_analysis/citations (accessed 19 May 2023).

United Nations (2015), 'Sustainable Development', Available at: https://sdgs.un.org/goals (accessed 17 May 2023).

Vezzoli, C., Conti, G.M., Macrì. L. and Motta, M. (2022), *Designing sustainable clothing systems: The design for environmentally sustainable textile clothes and its Product-Service Systems*, Milano: FrancoAngeli s.r.l.

Wegener, C. (2016), 'Upcycling' in V.P. Glăveanu, L. Tanggaard and C. Wegener (eds), *Creativity – A new vocabulary*, 181–188, London: Palgrave Macmillan.

Chapter 3

Stitched Refugee Stories: Using Stitching and Upcycling to Convert Children's Experiences into Empowerment

Chanjuan Chen and Trina Harlow

Introduction

Refugee Crisis

Enormous numbers of internally and externally displaced people and forced migrants have been a significant global crisis and issue for many years. The numbers continue to rise at an alarming rate. According to the United Nations High Commissioner for Refugees (UNHCR), nearly 109 million people were forcibly displaced by the end of 2022. Over 41% of these were children under the age of 18 (UNHCR, 2023). Children are the most vulnerable population and have many unique challenges. There are many personal accounts of the challenges that refugees face. The documentary film titled *Refuge in the Heartland* shared the story of a middle-school paraprofessional currently working in the USA who worked with the Newcomer Programme students at her school in 2018. When she was a small child, she had to walk from Cambodia to Vietnam with only one mat, some dishes, and clothes (Kansas State University, 2018). Some parents must or choose to send their children away unattended. An American woman tells the story of her mother, who was put on a boat as an eight-year-old girl, alone and with only the clothes she was wearing, to flee from the Vietnam War. She had to travel by herself for several days on the sea to a refugee camp in Malaysia (Pham, 2020).

To provide immediate protection and help to refugees, the UNHCR has built refugee camps around the world. These camps are meant to provide basic needs to refugees (UNHCR, 2023). However, some of the refugees are forced to live in informal tents for extended periods. Tent cities can be found in countries such as France and the border between Mexico and the USA (Byrne, 2021; Lithwick, 2019). Some of these tent cities have poor living conditions caused by health, social, political, or environmental problems. These conditions are putting refugees, especially children, in danger (Lithwick, 2019). Moreover, though refugee camps are intended to serve as temporary facilities providing protection and support, many children have lived in refugee camps from birth to adulthood and have never seen the outside world (Harlow, 2019).

Researchers found that education is one way to change refugee children's lives (Huss et al., 2020; Piper et al., 2020). Education is a life-saving, life-sustaining, and life-transforming process that forms the basis of children's futures, providing them with the understanding, knowledge, and skills to create more prosperous lives for themselves and their communities (Shohel, 2022). How can we transform downtrodden and traumazsed refugee children into empowered individuals through quality education and other resources in refugee camps, promoting their growth and development? In other words, how can we upcycle the educational experience of refugee children, no matter their location in the world? To bring attention to the challenging educational and developmental story of refugee children, an art-based research approach using upcycling created a social justice fashion design to tell some of these refugee stories and inspire continued effort in this field.

Upcycling

Refugee education and fashion seem to collide in a reimagining of how to improve the status quo. Fast and mass fashion have created an overwhelming amount of consumption and waste in our world today (Bianchi & Britwistle, 2011; Fletcher, 2010; Niinimäki & Koskinen, 2011). An increasing range of low-quality clothing is rapidly disposed of, leading to higher consumption of inexpensive products and massive textile waste (Farrer, 2011). One way to reduce the amount of textile waste is to take materials that would otherwise be thrown away and reuse, recycle, and upcycle. Reusing existing materials could provide designers with the opportunity to reassess the value of waste material (Gwilt & Rissanen, 2011). Upcycling is defined as recreating a product that has higher quality than the original. Many fashion researchers and practitioners have explored upcycling as a design method. For example, in the work of fashion designer Cloed Priscilla Baumgartner, men's trousers were transformed into thoroughly modern pieces for women under the label "MILCH" (Brown, 2013). Fashion Researcher (2017) produced a collection of upcycled garments presenting various ways of reusing and adding value to wasted t-shirts, men's dress shirts, and fabric scraps. Irick (2013) proposed a model for repurposing apparel and textiles. The model included three levels of repurposing: redecorating, subtractive and additive. A fourth level of repurposing to this model, intentional pattern making, was added by Irick and Eike (2017) in the later study. These four levels provided a clear guideline for teaching students about the method of repurposing and upcycling in the classrooms.

On the other hand, Wegener (2016) argued that upcycling is not just a design approach but a way to link the past to the present and the future; here we see another parallel with the examination of refugee education. In upcycling, the present incorporates elements from the past; the future becomes tangible, enabling individuals to establish connections between the significance of both old and new items. As defined by Akdemir (2018), fashion is more than a piece of fabric; it is 'a form of non-verbal communication' (p. 1391) and 'one of the most important ways of expressing the identity because of its full of signs, codes and meanings inside them' (p. 1396). Researchers in textile and apparel have explored fashion that tackles social issues (An et al., 2020; Hahn & Collier, 2022). However, few apparel and textile researchers have explored how to use upcycled materials to create garments that bridge the meanings of the old materials with the new garments while addressing social issues to promote positive change.

In this chapter, we describe an Arts-Based Research project (Cahnmann-Taylor, 2008) using recycled materials to tell the challenging stories of refugee children. By analysing refugee statements and photographs from selected research on refugee children's education, we upcycled a used tent to transform the meaningful finding into a new fashion piece. The concept of upcycling aligns with the purpose of this research, which is to bring attention to educational challenges that affect refugee children and to the value of each individual through quality education and resources. Therefore, the purposes of this study are twofold: (i) to explore innovative ways of creating a fashion piece using upcycled and unconventional materials; and (ii) to provide insight, knowledge and attention, allowing deeper understanding of and support for the educational opportunities for refugee children.

Approaches and Literature Review

The methodological framework for this study is Arts-Based Research (ABR), which implements various visual art forms to collect and analyse data and present findings (Cahnmann-Taylor, 2008). ABR is not only rigorous, relevant and insightful but also continues to explore new dimensions at the scientific frontier (ibid). ABR allows researchers to explore unconventional paths, producing outcomes that surpass the constraints of traditional sciences (Rolling, 2010). ABR guided this study by affirming that data can be converted into visual designs. To collect the data, we analysed and synthesized literature on educational research with refugee children and visual images associated with these articles. Examining images allowed us to discover new nuances and subtleties in situations, leading to a deeper understanding and expanding our awareness (Eisner, 2008). We collected and reviewed a total of 12 peer-reviewed articles (Adams & Santos, 2022; Ardelean, 2021; Brown & Bousalis, 2017; Crea et al., 2022; Demissie & Boru, 2021; Heldal et al., 2021; Huss et al., 2020; Kodwani, 2019; Meyer et al., 2013; Piper et al., 2020; Shohel, 2022; Wellman & Bey, 2015), two policy reports (Dryden-Peterson, 2015; Palik, 2020), and one book (Harlow, 2019) from Google Scholar and academic publisher databases, as well as articles and images on national agencies and organizations that protect and support refugee children, such as the United Nations High Commissioner for Refugees (UNHCR), U.S. Committee for Refugees and Immigrants (USCRI), and United Nations International Children's Emergency Fund (UNICEF), providing evidence-based research on the experiences and education of refugee children. The search was conducted using keywords such as 'refugee children', 'refugee education' and refugee camps'.

By analysing the literature and visual images, we identified several key points regarding the challenges of education for refugee children and the pedagogical strategies proposed by researchers. These key points informed the authors' decision to act on experimenting with different apparel and textile techniques, such as upcycling and embroidery. Consequently, an upcycled fashion piece was created, incorporating quotes and images from the literature review as virtual surface details on the garments.

Analysis of Literature and Visual Images

A review of the literature and visual images on the topic of refugee children and education points to two essential issues: (i) the educational challenges and (ii) the pedagogical strategies needed for refugee children. Studies pertinent to each are discussed in this section to provide the foundation for the development of wearable art to promote awareness and support for the refugee educational crisis. A summary of the literature review map was developed based on the existing research and used in support of the second-hand materials selection and the upcycled garment development.

Through a review of the literature, some challenges of educating refugee children have been discussed. As Piper et al. (2020) found, one of the challenges was related to language. The language(s) spoken by refugee students at home and in their previous schooling can significantly impact their learning outcomes after being displaced. The research proposed that providing effective educational opportunities for refugee children should consider these factors when deciding on the language of instruction (Piper et al., 2020). A different challenge pointed out by Demissie and Boru's (2021) study surrounded the choice of the curriculum offered for refugee children in the host country. The study highlights two conflicting viewpoints in the literature. One advocates that refugees should adopt the educational curriculum of the host country, while the other proposes that they follow the curriculum of their country of origin. In their conclusion, Demissie and Boru (2021) proposed a middle ground by suggesting the implementation of a negotiated curriculum, which resulted from a mediation process between the home and the host country's curricula. This approach aimed to provide better support for the learning of refugee children. The challenges related to diverse materials and languages correspond with those found in Shohel's (2022) study on educating Rohingya children in Bangladeshi refugee camps. Shohel's research exposed government limitations on using the Bangladeshi curriculum and Bangla language for instruction. This led to challenges in obtaining affordable and suitable learning materials to provide quality basic education in the children's mother tongue in 2022.

To provide a quality educational environment for refugee children, many researchers have proposed different pedagogical strategies through studies around the world. For example, Huss et al. (2020) explored educational strategies to support learning in a refugee camp in Mitilini, Greece, such as using art-based methods to understand children's needs and create a conducive learning environment. The authors emphasized the importance of formal learning for refugee children as a source of hope as well as the importance of refugee teachers as they became role models for the children. Art lesson activities and community projects empower young minds, foster creative expression and reconnect refugees with their inherent sense of self-worth and belonging (Brown & Bousalis, 2017).

For the review of photographs and images, we analysed evocation to examine the visual images, focusing on the initial emotional responses when viewing each image (Saldana & Omasta, 2022). The majority of the visual images were sourced from the literature analysed above, with additional images found through a Google search using the exact keywords used in the literature review process. For example, we reviewed one drawing from Huss et al.'s (2020) study to get an idea of what school looks like from the perspective of a refugee child. When we examined the drawing, our eyes naturally gravitated toward the colourful house in the middle of the drawing. We jotted down the thoughts and emotions first evoked within us: colourful, positive and hope. A total of nine visual images were reviewed and analysed using the same method.

In analysing the literature and visual images, key points were identified and a literature map was developed, highlighting the important dimension of educational challenges and strategies for refugee children (Figure 4.1). The literature map informed the development of Arts-Based Research. Upcycling was selected as the design technique as the concept of upcycling aligns with the purpose of this chapter, which is to help refugee children reassess the value of each individual through quality education and resources in the new country. Additionally, as fashion has the power to tell stories, an upcycled fashion piece was viewed as an appropriate visual form to explore the research further.

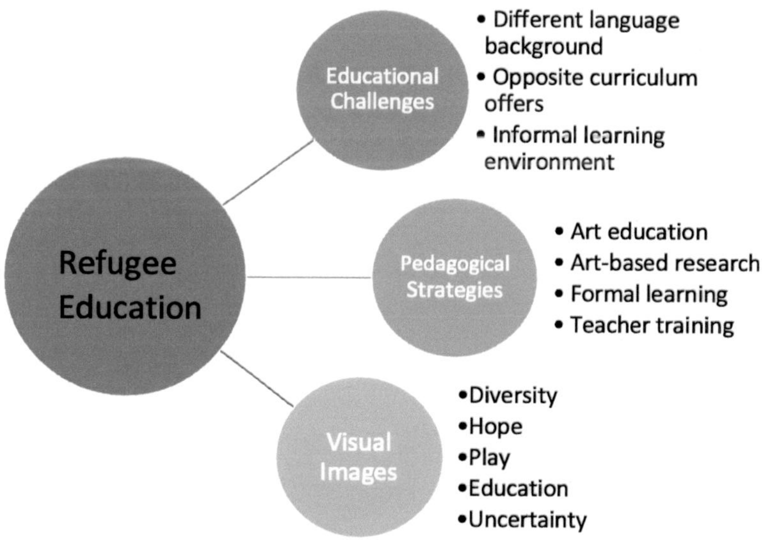

Figure 3.1 Literature map with key points.

Fashion Techniques

The upcycling design process began with finding discarded materials for this project. Since the inspiration behind this fashion piece was the global refugee crisis and the educational challenges faced by refugee children, a second-hand dome tent was seen as an appropriate material due to its frequent use by refugees on their journey and its colours and textures aligning with the finding. The tent had a grey and blue groundsheet made from tarp. The tarp is textured with a slightly rough surface that hints at its durability. A subtle graininess, like refined grains of sand, can be felt when running one's fingers over the tarp. The texture and the feeling of the tarp symbolize the challenges of the refugee children. The body of the tent, made from waterproof polyester, was mixed with a blue-and-white colour. The vents on the tent were made from blue-and-grey mesh materials. The colour palette of the tent, grey, white and blue, creates a profound narrative that embodies the stories of refugee children – the grey signifies the challenges discussed in the literature, the white symbolizes a blank canvas where their new life's potential in a new country is yet to unfold. The blue is like the blue sky, which inspires a sense of hope and wonder about what lies beyond. The initial design process involved sketching and draping the repurposed tent fabric on dress forms to explore potential silhouettes. After careful consideration, the authors selected a simple full-length dress with a hooded cape as the final design. This decision was guided by both practical and aesthetic considerations. The elongated silhouette, featuring a long train, efficiently utilized the tent's body without generating additional waste while also providing ample space for embroidery. Additionally, the hooded cape was incorporated to symbolize protection, leveraging the inherent properties of the tarp material to reinforce this conceptual narrative.

Different fashion techniques were explored for this upcycling process, such as modularity and embroidery. Modular design is a design approach that features small, standardized units that can be independently combined in various configurations to create different forms (Koo et al., 2013). In this study, modularity refers to the uncertainty of the refugee children's future. The embroidery technique allows for intricate visual designs using thread, serving as an artistic means to secure and enhance materials of various types and sizes (Haar, 2004). The repurposing method known as intentional pattern making (Irick & Eike, 2017) was also incorporated. This method involves taking apart old clothes and creating new designs using the fabric available while using existing shapes and space efficiently. The intentional pattern-making method can maximize the characteristics of the old material in the upcycling process to bridge the meanings of the old materials with the new garments.

Outcomes

The Development of the Upcycled Garments

To create the upcycled fashion piece, the old dome tent was first cleaned and disassembled. The grey and blue tarp materials that formed the base of the tent were kept, creating a cape. The polyester materials from the body of the tent, which had a smooth texture and interesting construction details like zippers, buckles, and flat felled seams, were kept intact to create the dress. A modular system based on a polygon shape was employed to create the cape (Chen & Lapolla, 2021). Children's body-shaped silhouettes were designed in a polygon shape using Adobe Illustrator and arranged in a circular pattern to symbolize a scene of children studying in school and to symbolize the refugee children that are the focus of this design. Functional cut-outs were added to each of the six body outlines. A laser cutter was used to cut out the modules on the grey and blue tarp materials (Figure 4.2). These modular shapes were then joined together through the cut-outs to form the cape, with smaller shapes on the neck and gradually larger shapes on the hood and sleeve hems. Modules made from blue polyester mesh, taken from the body of the tent, were also added to enhance the visual effect of the gradient colour palette on the cape.

For the dress design, unconventional shape and construction details of the dome tent were used instead of dismantling the tent, the whole body of the tent was draped on a dress form in various ways to test design ideas. A floor-length dress with a train was eventually chosen as the silhouette. The draping process involved analysing the available materials and existing construction details from the tent to preserve its characteristics and achieve the desired new look. For example, the zipper from the tent opening was utilized as the closure of the dress on the centre back. Next, embroidery details using a free motion machine were added to the dress, incorporating direct quotes from nine selected publications on educational research in refugee camps. The authors also developed outline drawings based on visual images associated with the literature to capture the essence of these findings. Both the quotes and the drawings were printed on sheets of wash-away embroidery stabilizer before stitching using a quilting machine on the front of the dress. Additionally, the names of the 10 largest refugee camps in the world were stitched on the back train. Once the desired stitching was added,

Figure 3.2 Children's body-shaped module design to be used on laser cutting on the grey tarp materials.

the stabilizer was washed away. The dress was then sewn together with an upcycled lining fabric to complete the look.

Merge of Social Justice and Upcycling

The final ensemble consists of two pieces created from a second-hand dome tent adorned with embroidery details (Figure 2.3). The modular cape is crafted using the grey-and-blue tarp sections of the tent. The colour palette on the cape transitions from grey to blue, symbolizing the current situation of refugee children (grey) and their aspirations for a better future through education (blue) in a new country. The cape features modules in the shape of children's silhouettes surrounding a polygon, representing a scene of studying. The rough texture of the tarp materials used for the modules signifies the hardships refugee children endured on their migratory journey. As for the dress, it was made from the body portions, the black-and-white polyester of the tent, accentuated with black embroidery details to complement the cape. The embroidery details included quotes from published scientific literature supporting refugee children's education and drawings developed by the researcher (Figure 2.3).

Arts present a multitude of opportunities and encourage educational researchers to think innovatively about the definition of research, to explore diverse and imaginative methods for empirical inquiry, and to communicate our inquiries and discoveries in more impactful and accessible ways (Cahnmann-Taylor, 2008). Through creating an upcycled fashion piece, we explored not only innovative ways to use upcycling as a design technique but also merging meaningful findings on the topic of refugee children with upcycling. The modular cape conveys

Figure 3.3 The front and back views of the upcycled cape and dress.

meaningful messages using shapes, colours and textures derived from upcycled tent materials. While a cape design was used, the modules could be transformed into different looks, symbolizing the unpredictable education environment in refugee camps. In combination with the dress, direct quotes and representative drawings embroidered on the dress depict the educational experiences of refugee children in camps. The two-piece fashion ensemble serves to amplify authentic voices and research knowledge from refugee camps worldwide. Further, the references of the publications were digitally printed on the backside of the stripes hanging from the cape. Through the stories of refugee children and the techniques of modularity and embroidery, creative upcycling allows the connection between the past stories of individuals and their new significant future.

Discussion

With UNHCR (2023) and other agencies that focus on the world refugee crisis reporting on the alarming increase in displaced people, the plight of the children and adults affected by the refugee crisis is a world issue and a concern to the global population. This chapter touched on an overview of the many challenges facing the education of displaced children. The goal of the research was to merge fashion with a global social justice issue and uniquely incorporate the refugee crisis into the literal fibres and stitches of a garment focused on the refugee child.

The upcycled dress featured as a part of this study functions as a soft goods billboard for the world refugee crisis. Therefore, the garment and its analysis serve as a challenge for upcycling designers and other fashion designers to incorporate social justice more frequently and overtly into designs. During recent Paris Fashion Shows, we have seen an uptick in the social focus of garments, with the themes of designs reflecting the concerns of individual designers (Web, 2022). This study indicates that there is much more that can be done regarding upcycling and refugees. Designers representing a variety of organizations could host a design course where refugee camp residents upcycle various fibre items (e.g., clothing, bedding, tents) and refashion them into clothing. Then, a fashion show could be held at the camp, and residents of the camp could serve as models. Additionally, castaway garments found and collected along the Texas and Arizona borderland could be upcycled into social justice garments that tell the stories of those border crossers.

The upcycled tent dress and fashion design of this chapter indicate that the voices of our refugees and their stories can live on through many venues and voices. More specifically, this chapter and its visual artefact – the upcycled dress – indicate that the design industry can do much more to join the global conversation on the crisis of displaced people. Words and images can only be considered when a person is reading or viewing them – a dress can bring great attention and notoriety to a cause, as was seen in 2021 when Alexandria Ocasio-Cortez wore an attention-getting dress to the Met Gala in New York City (Hills, 2021). The fashion design of the future will be more than just fashion.

Conclusion

This chapter synthesized refugee educational research and creatively addressed educational challenges for displaced and refugee children through the production of an upcycled garment. The visual design serves as a call to action and empowerment, drawing attention to the issues faced by refugee children to access education. This fashion piece ventures into innovative approaches to create upcycled modular designs while addressing educational challenges in refugee camps. The findings conveyed the journey of real experiences and research from refugee camps worldwide, offering valuable insights to support the education of refugee children during their transition to new communities. Additionally, the method employed to upcycle the tent contributes to the existing knowledge of intentional pattern making (Irick & Eike, 2017) by exploring unconventional materials and reusing construction details for new designs. The cumulative effort of the fashion piece also added a new aspect of how upcycling can be used to link the past value of the garment with the new art form created using various fashion techniques. Future studies will further explore additional methods of creating modular wearable art pieces addressing social issues, with the goal of advancing a new design framework for the fashion design industry and designers.

References

Adams, N. B. and Santos, R. M. (2021), 'Resettlement education case managers' roles supporting refugee families with children with disabilities'. *The Journal of Special Education*, 56(2): 63–72.

Akdemir, N. (2018), 'Visible expression of social identity: the clothing and fashion', *Gaziantep University Journal of Social Sciences*, 17(4): 1389–1397.

An, S. K., Lee, H. J., May, H. and Zainea, N. (2020), 'As my name', *International Textile and Apparel Association Annual Conference Proceedings*, 77(1): 1–4.

Ardelean, A. (2021), 'Play in a refugee camp: disorder from chaos', *International Journal of Play*, 10(4): 355–360.

Bianchi, C. and Britwistle, G. (2011), 'Consumer clothing disposable behaviour: A comparative study', *International Journal of Consumer Studies*, 36(3): 335–341.

Brown, S. (2013), *Refashioned: Cutting-edge clothing from upcycled materials*. London: Laurence King Publishing.

Brown, S.L. and Bousalis, R. (2017), 'Empowering young minds through communication, creative expression, and human rights in refugee art', *Art Education*, 70(4): 48–50.

Byrne, M. (2021), 'On the streets of Paris: the experience of displaced migrants and refugee', *Social Sciences*, 10(4): 130.

Cahnmann-Taylor, M. (2008), 'Arts-based research: Histories and new directions', in M. Cahnmann-Taylor and R. Siegesmund (eds), *Arts-based research for education: foundations for practice*, 21–33, New York: Routledge.

Chen, C. and Lapolla, K. (2021), 'The exploration of the modular system in textile and apparel design', *Clothing and Textiles Research Journal*, 39(1): 39–54.

Crea, T. M., Klein, E. K., Okunoren, O., Jimenez, M. P., Arnold, G. S., Kirior, T., Velandria, E. and Bruni, D. (2022), 'Inclusive education in a refugee camp for children with disabilities: How are school setting and children's behavioral functioning related?' *Conflict and Health*, 16(1): Article 53.

Demissie, E. A. and Boru, A. K. (2021), 'Curriculum choice for refugee children at Jawi Refugee Camp, Ethiopia', *Transnational Curriculum Inquiry*, 18(1): 67–77.

Dryden-Peterson, S. (2015), *The educational experiences of refugee children in countries of first asylum*. Migration Policy Institute.

Eisner, E. (2008), 'Art and Knowledge' in J.G. Knowles and A.L. Cole (eds), *Handbook of the arts in qualitative research: perspectives, methodologies, examples, and issues*, 3–12, London: Sage Publications.

Farrer, J. (2011), 'Remediation: discussing fashion textiles and sustainability', in A. Gwilt and T. Rissanen (eds), *Shaping sustainable fashion: changing the way we make and use clothing*, 19–34, Washington, DC: Earthscan.

Fletcher, K. (2010), 'Slow fashion: an invitation for systems changes', *Fashion Practice*, 2(2): 259–265.

Gwilt, A. and Rissanen, T. (2011), *Shaping sustainable fashion: Changing the way we make and use clothes*. London: Earthscan.

Haar, S. J. (2004), 'Wearable art through free-motion stitching', *Clothing and Textiles Research Journal*, 22(1–2): 31–37.

Hahn, K. H. and Collier, A. (2022), 'The chrysalis shroud for transgender', *International Textile and Apparel Association Annual Conference Proceedings*, 78(1): 1–5.

Harlow, T. D. (2019), *Journey to refuge: understanding refugees, exploring trauma, and best practices for newcomers and schools*, Manhattan: New Prairie Press.

Heldal, M., Hagen, T. L., Olaussen, I. O. and Haugen, G. M. (2021), 'Social sustainable education in a refugee camp'. *Sustainability*, 13(7): Article 3925.

Hills, M.C. (2021), 'AOC caused a stir with her statement-making Met Gala gown', *CNN Style*, 14 September. Available at: https://www.cnn.com/style/article/aoc-met-gala-alexandria-ocasio-cortez-dress/index.html

Huss, E., Asher, S. B., Shahar, E., Walden, T. and Sagy, S. (2020), 'Creating places, relationships, and education for refugee children in camps: Lessons learnt from the 'the school of peace' educational model', *Children & Society*, 35(4): 481–502.

Irick, E. (2013), *Examination of the design process of repurposed apparel and accessories: An application of diffusion of innovations theory*, Oklahoma: Oklahoma State University.

Irick, E. and Eike, R. J. (2017), 'Teaching the repurposing mindset: The introduction of a repurposing project into an advanced apparel construction course', *International Federation for Home Economics Conference Proceedings*, 23–25 March. Sligo: St. Angela's College.

Kansas State University. (2018), 'Refuge in the heartland: refugee resettlement documentary', Available at: https://coe.k-state.edu/about/documentaries/refuge-in-the-heartland.html (accessed 15 July 2023).

Kodwani, D. (2019), 'Living in a refugee camp and making way in life through education'. *Widening Participation and Lifelong Learning*, 21(2): 20–23.

Koo, H. S., Dunne, L. and Bye. E. (2013), 'Design functions in transformable garments for sustainability', *International Journal of Fashion Design, Technology, and Education*, 7(1): 10–20.

Lithwick, D. (2019), 'Trump's Tent Cities Are on the Verge of Killing Immigrant Children', *SLATE*, 23 December. Available at: https://slate.com/news-and-politics/2019/12/trump-tent-cities-mpp-killing-immigrant-children.html (accessed 29 July 2023).

Meyer, S., Murray, L. K., Puffer, E. S., Larsen, J. and Bolton, P. (2013), 'The nature and impact of chronic stressors on refugee children in Ban Mai Nai Soi Camp, Thailand'. *Global Public Health*, 8(9): 1027–1047.

Niinimäki, K., Peters, G., Dahlbo, H., Perry, P., Rissanen, T. and Gwilt, A., (2020), 'The environmental price of fast fashion', *Nature Reviews Earth & Environment*, 1(4): 189–200.

Ohrn-McDaniel, L. (2017), 'From 0 to 100: An exploration of waste in fashion', *ECO Fashion*, Available at: https://static1.squarespace.com/static/55072d6be4b03d6953f282e0/t/58fe3de13e00becefbc47914/1493057039600/catalogcolorado4.pdf

Palik, J. (2020), *Education for Rohingya refugee children in Malaysia*, PRIO Policy Brief.

Pham, S. (2020), 'A poem for my mother, a Vietnam boat refugee', *KUOW - NPR Network*, 22 October. Available at: https://www.kuow.org/stories/a-poem-for-my-mother-a-vietnam-boat-refugee

Piper, B., Dryden-Peterson, S., Chopra, V., Reddick, C. and Oyanga, A. (2020), 'Are refugee children learning? Early grade literacy in a refugee camp in Kenya', *Journal on Education in Emergencies*, 5(2): 71–107.

Rolling, J. (2010), 'A paradigm analysis of arts-based research and implications for education', *Studies in Art Education*, 51(2): 102–114.

Saldana, J. and Omasta, M. (2021), *Qualitative research: analysing life*. 2nd edn. Newbury Park: Sage Publications.

Shohel, M. M. C. (2022), 'Education in emergencies: Challenges of providing education for Rohingya children living in refugee camps in Bangladesh', *Education Inquiry*, 13(1): 104–126.

UNHCR (2023), 'Refugee statistics', *USA for UNHCR, The UN Refugee Agency*. Available online: https://www.unrefugees.org/refugee-facts/statistics/ (accessed 21 July 2023).

Webb, B. (2022), 'The year fashion finally faced its social impact', *VOGUE Business*, 29 December. Available at: https://www.voguebusiness.com/sustainability/the-year-fashion-finally-faced-its-social-impact (accessed 29 July 2023).

Wegener, C. (2016), 'Upcycling' in V.P. Glăveanu, L. Tanggaard and C. Wegener (eds), *Creativity – A new vocabulary*, 181–188, London: Palgrave Macmillan.

Wellman, S. and Bey, S. (2015), 'Refugee children and art teacher training: promoting language, self-advocacy, and cultural preservation'. *Art Education*, 68(6): 36–44.

Chapter 4

Exploring Upcycling Practices in the Fashion Sector: A Case of Harare

Edlight Mutungwe, Walter Chipambwa and Lucia Sithole

Introduction

The fashion and textile businesses are among the most polluting industries in the world, mainly because of the volume of production. Several studies have shown that the textiles and clothing industry contributes significantly to environmental pollution, leading to global warming (Morgan & Birtwistle, 2009; Minney, 2011; European Commission, 2017; Hiller Connell & Kozar, 2017; Leal et al., 2019). In addition to the effects of extracting raw materials for manufacturing new fibres, textile waste has emerged as a significant issue for the industry – approximately 30 million tonnes of textile waste every year. Upcycling has become one of the sustainable concepts being promoted in the fashion and textile industry in recent years to counter the negative environmental impact of this sector. Upcycling reduces waste as discarded materials are remodelled or redesigned into something more useful with a higher value than the original.

Janigo and Wu (2015) opine that upcycling fashion could involve minimal modifications, such as adding embellishments or deconstructing and reconstructing a garment. Aus (2011) defines upcycling as a circular design approach where waste clothes are converted through innovative designs to create new fashion products. Upcycling provides environmental benefits (Egin, 2012; Pol, 2010; Santulli & Langella, 2013), economic benefits (Zhuo & Lavendis, 2014; Fraser, 2011), and social benefits through community engagements (Sung et al., 2014). The design process in upcycling is different from the usual process based on new materials. The general design process in fashion upcycling involves sourcing, cleaning, sorting, designing, and production (Han et al., 2015; Chuma et al., 2019). The product is designed based on the inherent properties of the waste garment, thus calling for a high level of creativity so that the final product can be of high aesthetic appeal to the consumers (Aus et al., 2021; Han et al., 2017).

Preloved clothing items in developed economies are often donated to charities and end up being sold in developing economies, resulting in a boom in the trade of these second-hand clothes. This has increased the total amount of textile and clothing waste in these developing economies as these second-hand items are becoming another source of clothing waste. Zimbabwe has not been spared from this challenge of textile waste production (Tsiko & Togarepi, 2012), with poor textile waste disposal practices such as burning being prevalent in urban areas, as well as the influx of second-hand clothes (Chipambwa et al., 2016). To counter the effects of the clothing industry on the environment, some upcycling fashion brands such as Fanfare label, Paneros, Ruby Moon, and Beyond Retro have taken the lead in addressing the problem. They have designed clothing items for sale using waste collected from various fashion and clothing production companies.

This chapter aims to establish the level of fashion upcycling awareness among fashion consumers in Harare, Zimbabwe, and promote fashion upcycling to reduce textile waste dumping. Harare is the capital city of Zimbabwe and is home to approximately 120 registered clothing manufacturing companies. It seeks to stimulate fashion upcycling practices in urban communities for environmental sustainability by determining the fashion upcycling level of awareness among fashion designers and consumers, as these urban centres have become pivotal in the fashion consumption and manufacturing chain. It also sought to establish the current fashion upcycling practices among fashion designers and manufacturers in Harare and explore challenges and possible solutions that could be implemented.

Methods

To understand the upcycling practices in the fashion industry, the study applied a mixed-methods research design. The goal of mixed-methods research is to create a more comprehensive understanding of an issue by integrating both quantitative and qualitative methodologies into a single study. Mixed methods help explain some data abnormalities and/or give a more thorough and in-depth view of the topic being studied (Dunning et al., 2008). The qualitative component of the study allowed the researchers to explore the ways in which fashion designers and garment manufacturers, as fashion experts, participate in the construction of an upcycling model as part of their practice. Before their interviews, the selected expert participants were given access to various images from a study conducted in 2019 by researchers on upcycling with fashion design students. The interview had semi-structured questions and lasted for 45–60 minutes. All participants agreed to the request to record their interviews so that the researchers could transcribe them later. Thematic analysis of the interview transcriptions was aided by using NVivo software. A total of 10 experts (five fashion designers and five clothing manufacturers) were purposively sampled to participate in the interviews. To obtain industry experts, a letter of introduction was first sent to the management of the company, who then recommended an individual who they believed to be knowledgeable about the topic under study and also the overall research area. The quantitative component was an online survey that asked fashion consumers about their views and practices on fashion upcycling using Google Forms. The online survey link was shared via two popular social media platforms, Facebook and WhatsApp. Consent to participate in the survey was deemed granted upon receipt of the completed questionnaire. Internet Protocol (IP) address and email address were used to control for duplicate entries, and a question about the participant's city was controlled for the study area. The survey was closed after receiving fifty responses, which were structured questionnaires seeking to understand their views and practices on fashion upcycling.

Results

The ten expert interviews involved participants from the clothing industry and the individual fashion designers who agreed to take part as key expert informants in the study. The profiles of the interview participants are shown in Table 5.1. Sixty per cent of the participants were females, and all the participants had attained an educational qualification, with a diploma being the least qualified. Participants from clothing manufacturing companies held different positions, as shown by their job titles in Table 5.1.

Definition of Upcycling

To establish the experts' knowledge or perceptions of upcycling, the interviewees were asked what they understood by the term upcycling in relation to fashion or clothing design practice. Their definitions included recycling, reuse, patchwork, resource efficiency, and surface design (Table 5.2).

Upcycling Practice and Knowledge

Interviewees were asked to identify the upcycling practices they engage in or have engaged in as individuals or groups. They had some garments that they had

Table 4.1 Profile of expert interviewees

	Gender	Age	Exp*	Edu	Company profile/position
Participant 1-*P1*	F	32 years	4 years	1st Degree	Brand owner, fashion house
Participant 2-*P2*	F	34 years	6 years	1st Degree	Brand owner, fashion house
Participant 3-*P3*	F	42 years	15 years	Masters	Brand owner, fashion house
Participant 4-*P4*	M	35 years	8 years	1st Degree	Brand owner, fashion house
Participant 5-*P5*	F	40 years	13 years	HND	Brand owner, fashion house
Participant 6-*P6*	M	32 years	4 years	Diploma	Designer, clothing company
Participant 7-*P7*	M	36 years	10 years	Diploma	Production manager, clothing company
Participant 8-*P8*	M	33 years	7 years	Diploma	Production manager, clothing company
Participant 9-*P9*	F	36 years	10 years	1st Degree	Brand manager, clothing company
Participant 10-*P10*	F	39 years	6 years	1st Degree	Designer, clothing company

*Exp – Experience in area, Edu – Highest educational qualification, HND – Higher National Diploma

designed from unwanted clothes, and the key motivating factor was trying out new design ideas and also working around worn-out or spoiled garment areas in creating new styles. Three of the designers mentioned that they usually prefer working on or altering those oversized shirts they get from second-hand dealers to make new items like bum shorts or mini-skirts. One interviewee also mentioned that they make accessories like clutch bags to match the styles: 'I prefer reworking denim shirts from second-hand market to make shorts and floppy hats that people can wear during summer' (P3).

The interviewees mentioned that upcycling reduces the amount of waste that goes to landfills and helps to conserve natural resources. They also highlighted that upcycling encourages individuals and businesses to think creatively and find new and innovative ways to use materials that would otherwise be discarded. Upcycling can be a cost-effective way to create new products or materials. Another interviewee also stated that upcycling promotes the development of craft industries, as people without jobs or with limited access to employment are given the opportunity to participate in the economy (Table 5.3).

Evidence of upcycling practice and knowledge was more prominent amongst those participants who manage their brands. It was also emphasized that their activities are supported by the abundance of unsellable second-hand clothes and the need to turn unwanted clothes into better, more stylish products. One interviewee said, 'I get oversized and out-of-style second-hand clothes from the

Table 4.2 Upcycling definition

Theme	Data extracts
Recycling	'Upcycling is the process of recycling an old garment to make a new one' (P6)
Reuse	'... taking fabric used on from one garment to construct another new garment design' (P1)
Patchwork	'... process of using patches that are still in good condition to make a new garment' (P3)
Resource efficiency	'... making sure we maximize the use of all the fabric that is still in good condition to make another garment' (P5)
Surface design	'... adding more detail or decorations to a garment unlike its state from the shop' (P4)

Table 4.3 Upcycling knowledge

Theme	Data extracts
Creativity	'... process allows one to explore new design ideas that can result in new and distinct fashion products that are not replicable' (P1)
Reduce carbon footprint	'... using the fabric pieces repeatedly before they are discarded makes us reduce effects of landfill' (P4)
Income generation	'... unused trims or fabric pieces can be converted into useful items that can be sold and result in more income from such trims' (P9)
Craftwork	'... promotes craftwork and entrepreneurship development' (P6)

dealers in town when they select that which can be sold, and the other items graded as unattractive to the customers is then bundled and sold cheaply' (P2).

Challenges of Upcycling

During the interviews, the experts generally revealed that a lack of awareness of the circular economy is a major factor hindering the promotion of upcycling as a sustainable initiative. In one case, the interviewee stressed that unless designers see themselves as key contributors to a sustainable environment, principles such as upcycling will not be appreciated by the public. Upcycling requires the use of discarded or waste materials. The availability of these materials for designers is another challenge as it requires picking up the material and sorting it to come up with a good design. Another critical issue identified by the interviewees is the decision on the type of waste to be used – pre-consumer or post-consumer. One interviewee pointed out that the cost of making a garment becomes very high as there is no standard procedure to follow. The study also revealed that upcycling requires a high level of creativity to come up with a garment that can be sold on the market. Another interviewee also mentioned that the use of post-consumer waste can lead to a negative perception of the new upcycled product, as it may be seen as less valuable, making the sales process very difficult for the designer. Another one mentioned, 'The fashion design curriculum has to include upcycling as a sustainable design concept taught early in school so that designers are knowledgeable and start practising it in school' (P5). Jayasinghe et al. (2021) argues that a lack of skills and training can hinder the adoption of upcycling among fashion designers. The other factor highlighted by the interviewees was the issue of market demand, the demand for sustainable and eco-friendly fashion products that can influence the practice of upcycling. The interviewees claimed that if consumers are willing to pay more for sustainable fashion products, fashion designers may be more willing to adopt upcycling.

Fashion Consumer Survey Results

The survey was responded to by fifty people; 84% were female and 16% were male. The age group of 30–39 years had 44 per cent of respondents, followed by 20–29 years (28%), then 40–49 years (20%), and lastly, those above 50 years (8%). Regarding the frequency of clothing purchases, 40 per cent buy monthly, 34 per cent buy quarterly, 16 per cent buy half-yearly, and 10 per cent buy yearly. The respondents ranked their sources of new clothing items as follows: (i) fashion boutique stores (60%), (ii) second-hand clothing stalls (53%), (iii) bespoke fashion houses or designers (34%), and (iv) cash or credit fashion retail shops (14%). Sixty per cent of respondents showed that they understood what fashion upcycling was by choosing the best definition from those given, while 40 per cent defined it as recycling clothes. With regard to disposing of unwanted clothes, most respondents mentioned that they give them away to friends or relatives. In terms of purchase decisions between an upcycled garment and a non-upcycled garment (new), 85 per cent chose the new (non-upcycled) garment, and only 15 per cent chose the upcycled. Those who chose to buy the upcycled garments stated reasons such as promoting sustainability, supporting local designers, having exclusive design, saving money, and setting new trends or unique fashion styles. Some of the factors that affect fashion upcycling were listed as: (i) a lack of knowledge of sustainable fashion practices and their benefits, (ii) cultural issues that go against wearing second-hand clothes, (iii) negative perceptions as it is associated with being poor, (iv) fast fashion available, and (v) designers' lack of creativity. In order to promote upcycling as a practice, the respondents highlighted the need for education and awareness raising on how upcycling promotes a sustainable future, as well as support for small upcycling start-ups at a community level.

Discussion and Conclusion

One of the key issues revealed in the study was that interview participants defined upcycling as recycling even though they were different (Zhi, 2021). Another major challenge highlighted in the study was the issue of pricing and selling upcycled products. The interviewees explained that it can be challenging to achieve a higher price because customers assume that something made from unwanted material should have a lower value because it is seen as waste. However, this is not the case in many developed countries where sustainable fashion is highly valued. With new technologies now allowing for virtual fashion catwalk events (Chipambwa et al., 2022), fashion designers need to think about how they can market their brands globally and even develop these designs digitally (Chen et al., 2021). Another main issue was the lack of practical knowledge on upcycling in terms of how to create great garments through upcycling. Both interview and survey results showed that there is a need to integrate upcycling concepts into the standard fashion design curriculum so that it can be practised throughout the textile and apparel value chain. Governments also need to support fashion brands and designers who wish to promote sustainable behaviour through upcycling. Collaboration and networking among fashion designers can facilitate the adoption of upcycling practices, leading to the creation of new designs and innovative upcycling techniques.

In conclusion, exploring upcycling practices among fashion designers and consumers in Harare provided valuable insights into the potential for upcycling to promote sustainable development and ethical fashion practices in the fashion industry. Based on this research, we have made the following suggestions. We would like to see more collaborative fashion design projects on upcycling that can allow designers to share experiences and knowledge. Fashion designers in developing countries need to use their creative skills to expand upcycling practices through new innovative designs that can also be showcased and developed through digital platforms to generate income from multiple sources. We hope to see government incentives like tax rebates and awards to incentivize designers and other players in the textile and clothing value chain to embrace upcycling. Fashion designers need to be more creative in experimenting with different techniques, such as cutting, patching and embellishing, to transform old materials into new and exciting designs that can fetch higher value on the market. To improve the sourcing of materials, fashion designers should engage second-hand clothes dealers, local thrift shops, and textile waste centres. Working with local communities to set up small upcycling start-up projects to support the unemployed should provide a good platform to teach people basic sewing skills and build their confidence to participate in upcycling.

References

Aus, R. (2011), 'Trash to trend: Using upcycling in fashion design' [Unpublished doctoral dissertation]. Estonian Academy of Arts, Available at: https://www.digar.ee/arhiiv/en/books/15357.

Aus, R., Moora, H., Vihma, M., Unt, R., Kiisa, M. and Kapur, S. (2021), 'Designing for circular fashion: integrating upcycling into conventional garment manufacturing processes', *Fashion and Textiles*, 8: 1–18.

Bailey, K., Basu, A. and Sharma, S. (2022), 'The environmental impacts of fast fashion on water quality: A systematic review', *Water*, 14(7): 1073.

Chen, T., Yang, K. and Lee, Y. (2021), 'Development of virtual upcycling fashion design based on 3-dimensional digital clothing technology', *The Research Journal of the Costume Culture*, *29*(3): 374–387.

Chipambwa, W., Munyaka, C. and Mutungwe, E. (2022), 'Virtual exhibitions for final capstone fashion design project: A case of Chinhoyi University of Technology final year students', *International Journal of Costume and Fashion*, *22*(1): 25–35.

Chipambwa, W., Sithole, L. and Chisosa, D. F. (2016), 'Consumer perceptions towards second-hand undergarments in Zimbabwe: a case of Harare urban dwellers', *International Journal of Fashion Design, Technology, and Education*, *9*(3): 176–182.

Chuma, C., Muza, T. and Chipambwa, W. (2019), 'An examination of the upcycling design process of used clothes among fashion design students: Towards an approach for sustainable fashion design practice', *International Journal of Costume and Fashion*, *19*(2): 59–69.

Dunning, H., Williams, A., Abonyi, S. and Crooks, V. (2008), 'A Mixed Method Approach to Quality-of-Life Research: A Case Study Approach', *Social Indicators Research*, 85: 145–158

Eder-Hansen, J., Kryger, J., Morris, J., Sisco, C., Watson, D., Kiørboe, N., Petersen, S.D., Larsen, K.B. and Burchardi, I. (2012), 'The nice consumer: Toward a framework for sustainable fashion consumption in the EU', BSR, Copenhagen.

Emgin, B. (2012), 'Trashion: The return of the disposed', *Design issues*, *28*(1): 63–71.

Erdiaw-Kwasie, M. O., Abunyewah, M., Yusif, S. and Erdiaw-Kwasie, A. (2023), 'Does circular economy knowledge matter in sustainable service provision? A moderation analysis', *Journal of Cleaner Production*, *383*: 135429.

European Union (2017), 'Mapping Clothing Impact in Europe: The Environmental Cost.' Available at: http://www.ecap.eu.com/wp-content/uploads/2018/07/Mapping-clothing-impacts-in-Europe.pdf.

Han, S., Tyler, D. and Apeagyei, P. (2015), 'Upcycling as a design strategy for product lifetime optimisation and societal change', *Proceedings of the PLATE conference*, Nottingham: Nottingham Trent University.

Han, S. C., Henninger, C. E., Blanco-Velo, J., Apeagyei, P. and Tyler, D. J. (2017), 'The circular economy fashion communication canvas', *PLATE: Product Lifetimes and the Environment*, 161–165, Amsterdam: IOS Press.

Han, S. L., Chan, P. Y., Venkatraman, P., Apeagyei, P., Cassidy, T. and Tyler, D. J. (2017), 'Standard vs. upcycled fashion design and production', *Fashion Practice*, *9*(1): 69–94.

Hiller Connell, K. Y. and Kozar, J. M. (2017), 'Introduction to special issue on sustainability and the triple bottom line within the global clothing and textiles industry', *Fashion and Textiles*, *4*: 1–3.

Janigo, K.A. and Wu, J. (2015), 'Collaborative redesign of used clothes as a sustainable fashion solution and potential business opportunity', *Fashion Practice*, 7(1): 75–97.

Fraser, K. (2011), 'Case study: Upcycling materials for fashion', in A. Gwilt and T. Rissanen (eds), *Shaping sustainable fashion: Changing the way we make and use clothes*, 35–38, London: Routledge.

Kozlowski, A. and va Pesaran, D. M. (2021), 'Recontextualising and appropriating second-hand western fashion items in non-western contexts', in K. Sung, J. Singh, B. Bridgens (eds), *State-of-the-Art Upcycling Research and Practice*, 61–64, Cham: Springer.

Leal Filho, W., Ellams, D., Han, S., Tyler, D., Boiten, V. J., Paço, A., Moora, H. and Balogun, A. L. (2019), 'A review of the socio-economic advantages of textile recycling', *Journal of Cleaner Production*, *218*: 10–20.

McDonough, W. and Braungart, M. (2010), *Cradle to cradle: Remaking the way we make things*. Berkeley: North Point Press.

Minney, S. (2011), *Naked fashion: The new sustainable fashion revolution*, Oxford: New Internationalist.

Morgan, L. R. and Birtwistle, G. (2009), 'An investigation of young fashion consumers' disposal habits', *International Journal of Consumer Studies*, *33*(2): 190–198.

Niinimäki, K. (2017), 'Fashion in a circular economy', in C.E. Henninger, P.J. Alevizou, H. Goworek, D. Ryding (eds), 151–169, Cham: Palgrave Macmillan.

Nyavaya, K. (2021), 'Zimbabwe: Model Champions Repurposing of Waste Material', *The Standard*, 4 July. Available at: https://www.newsday.co.zw/thestandard/standard-people/article/13854/model-champions-repurposing-of-waste-material (accessed 16 January 2023).

Pol, V. G. (2010), 'Upcycling: converting waste plastics into paramagnetic, conducting, solid, pure carbon microspheres', *Environmental Science & Technology*, 44(*12*): 4753–4759.

RockstrÖm, J. (2009), 'A safe operating space for humanity Identifying', *Nature*, 461(7263): 472–475.

Runnel, A., Raihan, K., Castel, N., Oja, D. and Bhuiya, H. (2017), 'White Paper: Digitally Enhanced Circular Economy Within Global Fashion Supply Chains', *Reverse Resources*, 24 October. Available at: http://www.reverseresources.net/about/white-paper.

Santulli, C. and Langella, C. (2013), '"+ design–waste": a project for upcycling refuse using design tools', *International Journal of Sustainable Design*, *2*(2): 105–127.

Sung, K. Cooper, T. and Kettley, S. (2014), 'Individual upcycling practice: exploring the possible determinants of upcycling based on a literature review', *Sustainable Innovation 2014*, Copenhagen, 3–4 November.

Szaky, T. (2014), *Outsmart waste: The modern idea of garbage and how to think our way out of it*. Oakland: Berrett-Koehler Publishers.

Tsiko, R.G. and Togarepi, S. (2012), 'A situational analysis of waste management in Harare, Zimbabwe', *Journal of American Science*, 8(4): 692–706.
Zhi, Y. (2021), 'The Upcycling and reconstruction of garments and fabrics', *Art and Design Review*, *10*(1): 72–102.
Zhuo, C. and Levendis, Y. A. (2014), 'Upcycling waste plastics into carbon nanomaterials: A review', *Journal of Applied Polymer Science*, *131*(4): 39931.

Chapter 5

FABIO – MAKING JEANS SUSTAINWEARABLE

Pranoti Sonawane, Lovish Bandwal and Yagnesh Gohil

Introduction

The circular economy works as a system to ensure that materials do not become waste and that nature is regenerated by maintaining product and material cycles through practices such as maintenance, reuse, refurbishment, remanufacturing, recycling and composting (Ellen MacArthur Foundation, 2023a). The transition to a circular economy is still in its early stages, with only 8.6 per cent of resources currently being utilized in a circular manner (Papamichael et al., 2022). The fashion industry, with its widespread production of jeans, bears a heavy environmental burden, characterized by immense water consumption, air pollution and excessive textile waste (Peter & Mishra, 2023). The fashion industry's substantial environmental footprint is undeniable, highlighting the urgency for implementing circular approaches (Heshmati, 2017). Among the myriads of garments, jeans stand out as a global wardrobe staple with a profound environmental impact. Extensive research unveiled compelling statistics highlighting the staggering environmental impact of the jeans industry. As illustrated in Figure 6.1, a staggering 6 billion pairs of jeans are produced annually, yet a mere 15 per cent undergo recycling or upcycling. It takes around 65 litres of water to produce a single pair of jeans, contributing to critical water scarcity issues. Furthermore, the manufacturing process results in significant air and water pollution, as well as considerable waste and emissions (Webber, 2022; Levi Strauss & Co., 2023).

This study aims to implement a circular economy-based framework to revolutionize traditional jeans manufacturing by reducing waste, conserving resources and empowering stakeholders in the fashion industry. To find circular opportunities in the circular economy, work has been done to signify the technical and biological cycle of the lifecycle of a pair of jeans. The study unlocks a total of eight untapped ideas while finding circular opportunities through design thinking practices. This study introduces FABIO (named from the word Bio-Fabric) as part of the solution. It is an innovative service, emerging as a transformative force in the jeans industry, adopting sustainability principles and circular economy practices.

Figure 5.1 Facts on the manufacturing of jeans.

Approaches

Secondary research and qualitative methodologies were used. The secondary research proved instrumental in constructing a theoretical diagram and identifying circular opportunities, offering a structured foundation for the study. At the same time, qualitative research played a central role in the ideation phase of solution building, facilitating a nuanced exploration of perspectives and contributing to the richness and depth of the research findings. This dual methodology allowed the strengths of each approach to be leveraged, ensuring a comprehensive and insightful analysis of the research. The ideation process was conducted through co-creation discussions involving key stakeholders–farmers, designers, tailors, and customers. To validate and refine the proposed solution, we conducted stakeholder interviews with five tailors, four farmers, six designers, and ten consumers, incorporating their insights and expertise to ensure feasibility and relevance.

Comprehensive Research

At the outset, a thorough investigation into the life cycle of jeans was conducted with the help of a circular economy butterfly diagram (Ellen MacArthur Foundation, 2023b). This helped to unveil the intricate web of environmental challenges and the continuous flow of material inherent in conventional jeans manufacturing and lifecycle. This comprehensive research served as a foundation upon which the sustainability strategy would be built. Through a deep analysis of the environmental impact, resource consumption, and waste generation, we gained valuable insights into areas that demanded transformative change.

Circular Opportunities

The aim to embrace circular economy principles was brought to life through dedicated brainstorming and ideation sessions. Involving all stakeholders – farmers, designers, tailors and customers, these sessions were a fertile ground for creativity and new ideas. The goal was to explore innovative approaches that would allow for the recycling and upcycling of jeans, thereby disrupting the traditional linear model of fashion production and consumption.

Engaging Stakeholders

The mission towards sustainability and circularity cannot be realized in isolation. Therefore, a vital aspect of the research was to actively engage and involve stakeholders at every step of the journey. From farmers to tailors, designers and customers, the solution would seek to create a cohesive community committed to sustainable living. Stakeholder engagement sessions allowed for open dialogue, shared insights and the alignment of collective efforts towards a shared vision of environmental responsibility. Discussions with stakeholders focused on multiple aspects, such as the feasibility of using agricultural waste in fabric production, the expertise required to create modular designs, and the business viability of upcycling and repair services. Farmers provided insights into sustainable cotton cultivation and waste repurposing, while designers contributed expertise on modular garment construction to extend product life. Tailors played a crucial role in testing the adaptability of marked jeans for upcycling, ensuring the ease of transformation into new products. Meanwhile, customer feedback helped us refine usability, incentivization, preferences in sustainable fashion, barriers to adoption, and the effectiveness of return models.

Solution Design

With sustainability as a cornerstone, the team envisioned a transformative upcycling process that would breathe new life into discarded jeans and promote their integration into a circular economy. Drawing insights from sustainable practices and technologies, this solution aimed to minimize waste, optimize resources, and restore a closed-loop system where jeans could be endlessly regenerated and repurposed.

Upcycling Jeans in the Circular Economy

Butterfly Diagram

It is crucial to grasp how we can recycle materials once they are no longer in use. The butterfly diagram illustrates a system of continuous cycles in a value chain (Ellen MacArthur Foundation, 2023b). The butterfly diagram illustrates a loop that depicts the continuous circulation of materials within a circular economy. These cycles are divided into two parts: the *biological cycle*, which deals with biodegradable materials and the *technical cycle*, which deals with non-biodegradable materials. In the technical cycle, products and materials are continually used and remain in circulation so that they do not become waste. The biological cycle includes materials such as food or cotton clothing that can naturally return to the environment, creating a regenerative cycle. These materials can go back into the system positively and safely, helping to rebuild and restore the natural environment. In Figure 6.2, the technical and biological cycle of our jeans is illustrated, derived from insights gained through secondary research. This framework allowed for a detailed examination of the materials used in the jeans, exploring possibilities for transitioning from a linear to a circular model. By doing so, the materials were examined at a granular level, contemplating how to integrate them into a circular system. This diagram serves as a tool for envisioning and strategizing the approach to circular opportunities in the future.

Circular Opportunities

Through collaborative workshop sessions, the full potential of the circular opportunities identified within butterfly diagrams was harnessed, thereby initiating the ideation process. This led to the development of a solution, which was referred to as FABIOinspired by the amalgamation of 'Bio' and 'Fabric'. Following the analysis, the FABIO concept was designed with the objective of facilitating the seamless integration of innovative practices by strategically interconnecting stakeholders from farmers to designers and customers. This is illustrated in Figure 6.3. At its core, FABIO envisions a world where jeans are no longer seen as disposable items and are destined for countless lives through upcycling and regeneration. FABIO plays a pivotal role in the collection of food crop waste from farmers, a practice which has traditionally involved burning the waste, resulting in the emission of pollutants into the atmosphere. Now, this food crop waste will be utilized by the jeans manufacturing industry. FABIO's circular economy employs recycled water and bio-fertilizers to cultivate crops by reusing the water, subsequently converting food crop waste into a valuable resource. Additionally, FABIO gathers seeds and lint from the cotton, directing them towards the oil production sector rather than discarding them. The utilization of the Agraloop technology enables the transformation of cotton and food crop waste into biodegradable fabric forms, which serve as the foundation to produce upcycled jeans (Circular Systems, 2023). Upon reaching the end consumer, FABIO offers solutions for upcycling the jeans, encompassing do-it-yourself (DIY) approaches, tailoring services, and other creative transformations.

Upcycling Jeans

By providing jeans with coloured lines on their inner side, as shown in Figure 6.4, FABIO empowers customers to embark on a personal journey of upcycling. With each upcycled pair of jeans, FABIO paves the way for a more sustainable and vibrant fashion future that empowers, uplifts, and regenerates the value of both garments and the planet.

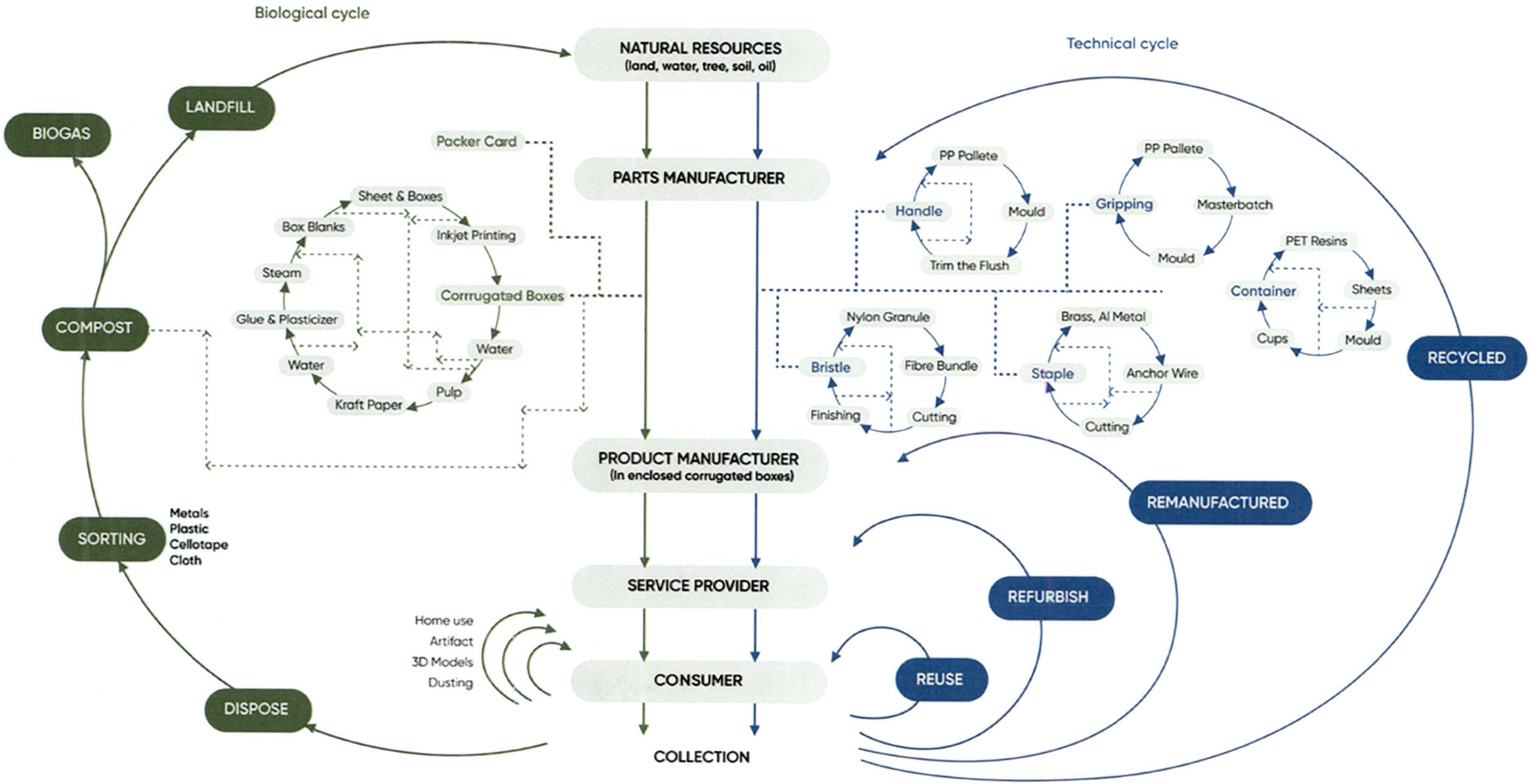

Figure 5.2 Butterfly diagram analysis of jeans.

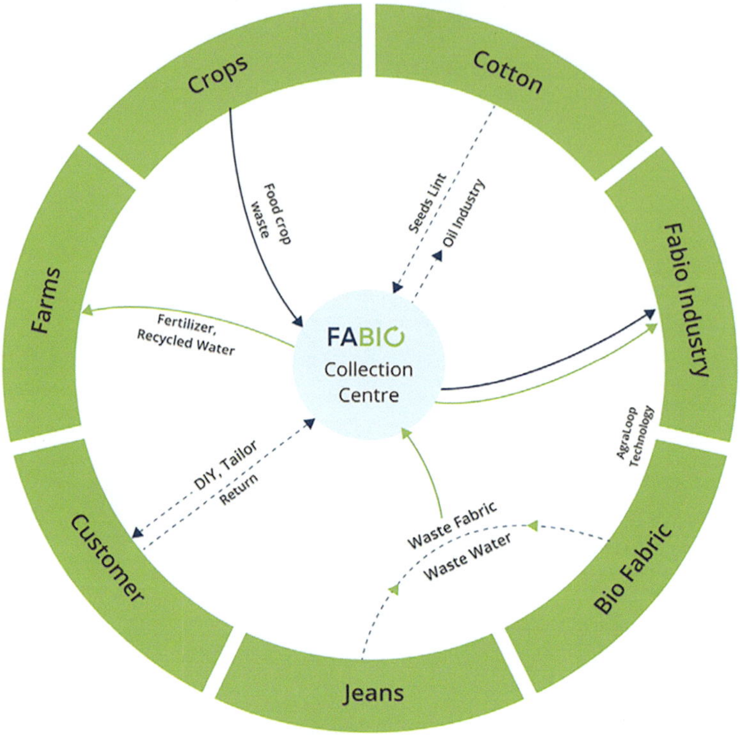

Figure 5.3 Circular opportunities of FABIO.

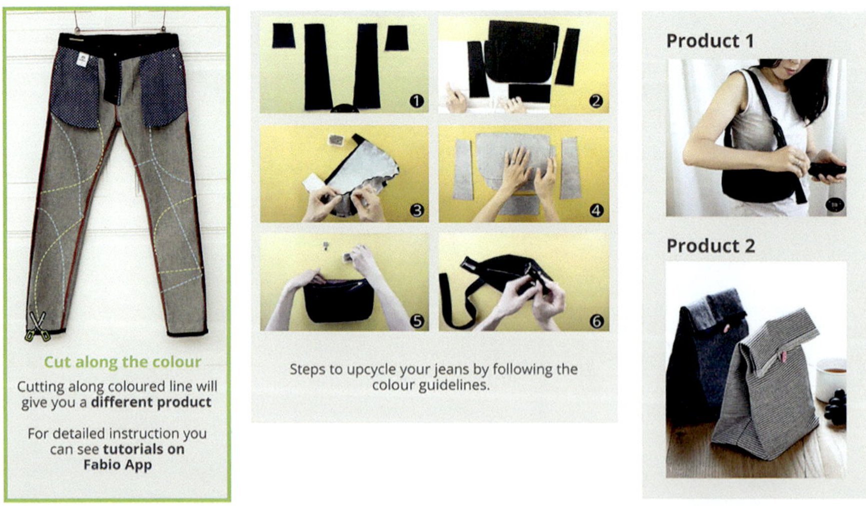

Figure 5.4 Coloured lines for customers to upcycle jeans on their own in different product styles.

FABIO Mobile Application

FABIO centres around implementing a circular economy, effectively addressing the environmental challenges associated with traditional jeans manufacturing. The integration of circular opportunities and sustainable practices has the potential to reduce waste, conserve resources and empower stakeholders (Negrete-Cardoso, 2022). Beyond environmental preservation, FABIO's unique upcycling solution fosters opportunities for tailors, farmers, designers, and the public, promoting sustainable practices throughout the fashion industry. Currently, the FABIO app is at the concept stage, with its framework and implementation strategies being explored through research and stakeholder engagement.

Do It Yourself and Tailor Support

Figure 6.5 shows the step-by-step process for customers to upcycle jeans through DIY. With simple cutting and stitching techniques along the marked lines, customers can transform their jeans into an array of new products, breathing new life into their wardrobe with each iteration. Through the FABIO app, customers can access tailored support for upcycling or venture into DIY projects, learning from tutorials to create unique products on their own. Such inclusivity not only empowers customers but also sustainably celebrates individuality and creativity.

Figure 6.6 shows the step-by-step process for customers which allows tailors to support customers directly. FABIO's engagement with local tailors has created a network of skilled artisans who collaborate with customers in upcycling projects.

User Interface

With the help of Tailor Support, customers can place an order and convert their old FABIO jeans into exciting products, as shown in Figure 6.7. Tutorials and DIY projects are available on the FABIO app to help customers get creative. They can order the materials shown in the videos to upcycle their product. Customers can use the return option to send back their old FABIO jeans. In return, they will receive reward points.

As shown in Figure 6.8, **u**sers can showcase their designs at FABIO Events, such as exhibitions and workshops. These designs can also be featured on the FABIO app, motivating users and enhancing their unique fashion brand image. Customers earn rewards and discounts for every purchase and return, helping the environment through reward points while also motivating them.

Functional Components of FABIO

The following are the functional components of FABIO:

- *Upcycling Made Easy*: The FABIO app empowers users to upcycle their old denim jeans effortlessly. With user-friendly video tutorials and step-by-step guidance, anyone can transform their jeans into exciting and stylish products.

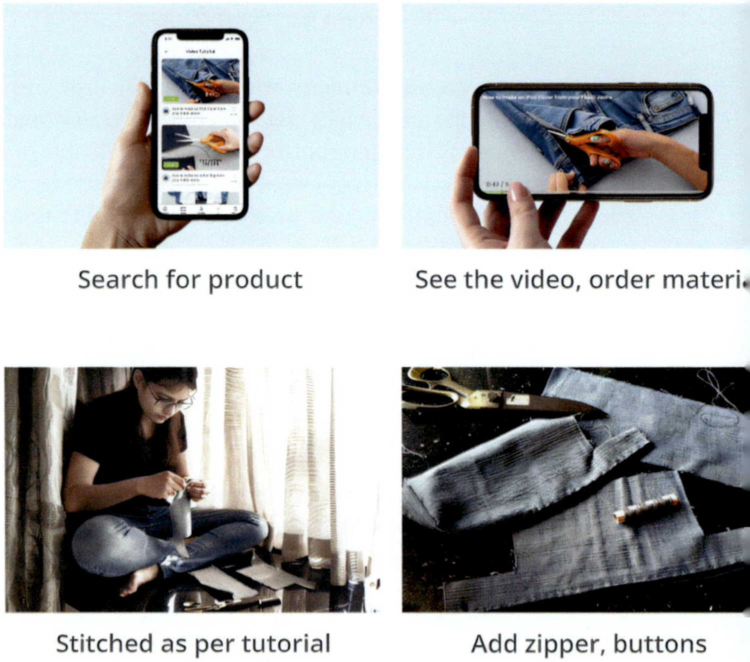

Figure 5.5 A step-by-step process for customers to upcycle the jeans through DIY.

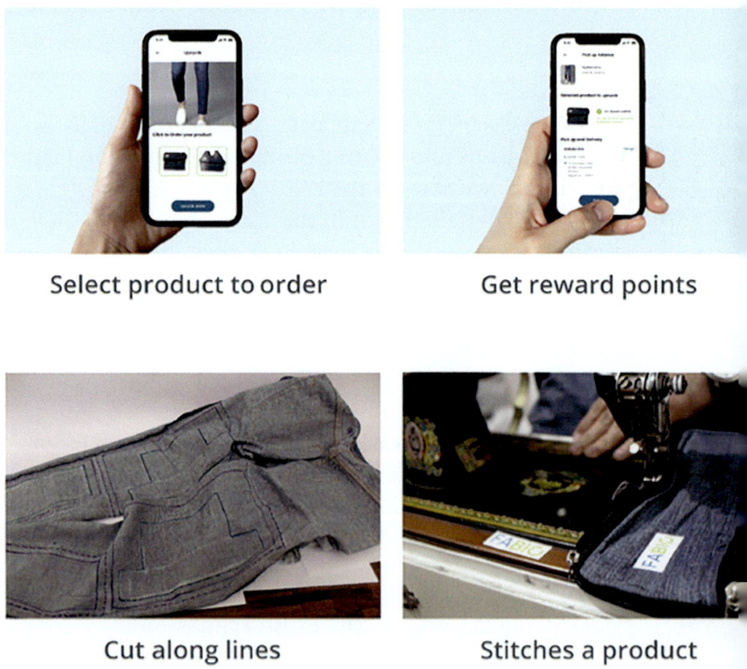

Figure 5.6 A step-by-step process for customers to take Tailor Support.

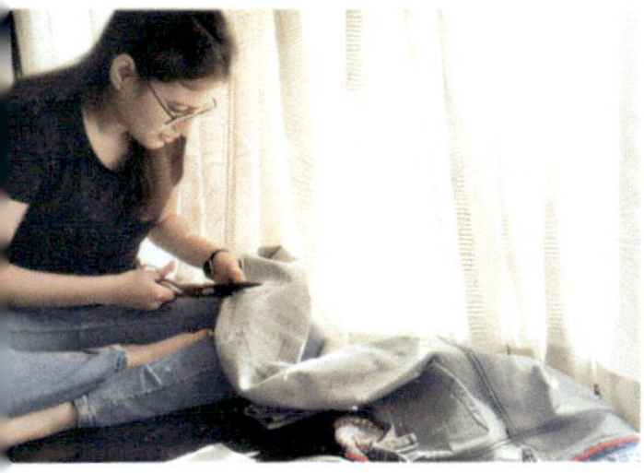

Cut the jeans

Patches ready to stitch

Turn over the sides

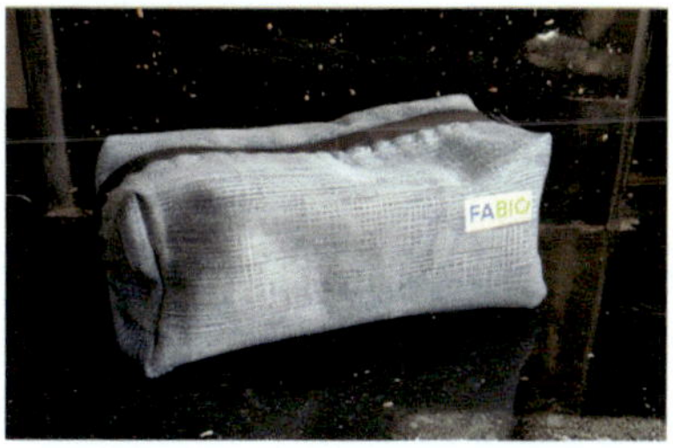

Upcycled and usable product

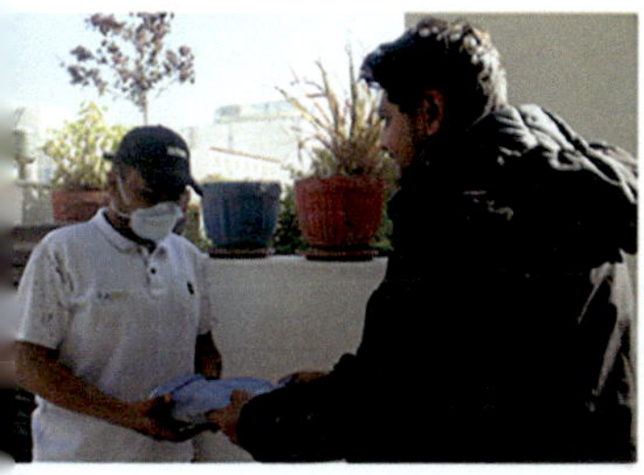

Jeans pickup

Delivery to tailor

Finished product

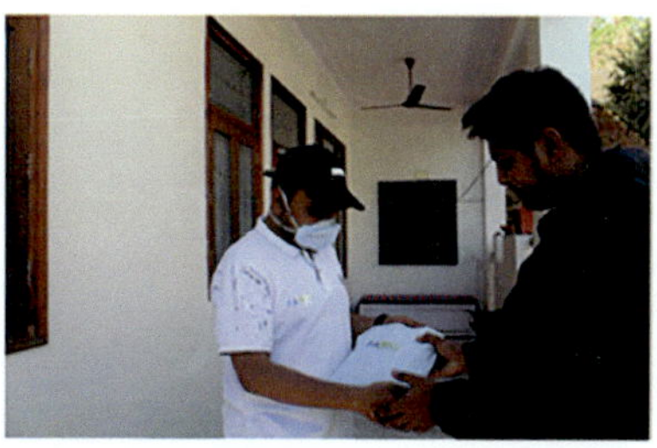

Upcycled product delivery

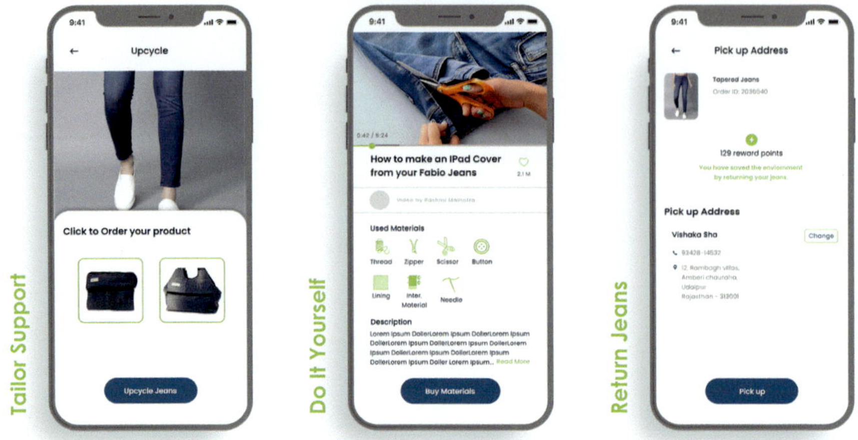

Figure 5.7 User interface for DIY and Tailor Support.

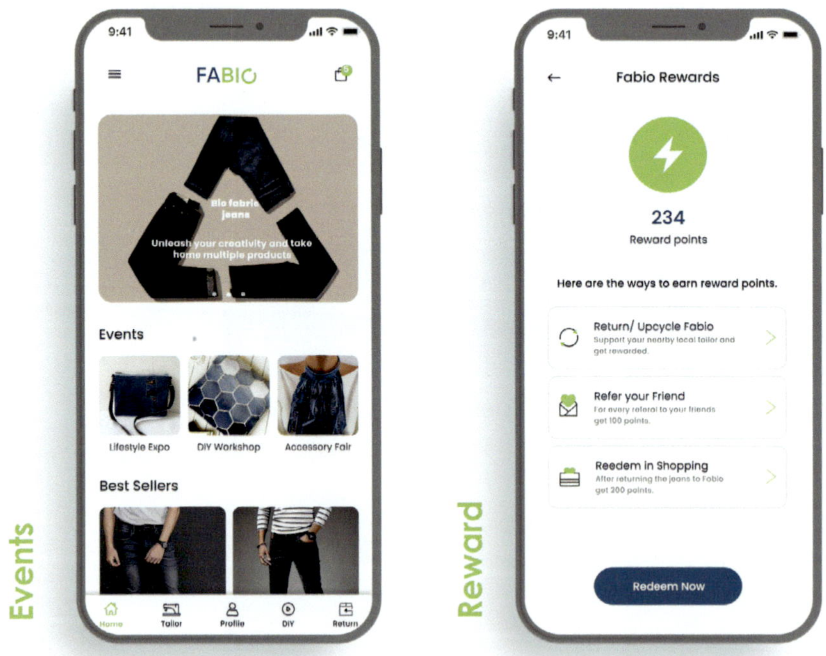

Figure 5.8 User interface for events and reward.

- *Support from Local Tailors:* For those who prefer professional assistance, the FABIO app connects users with local tailors specialized in denim upcycling. Tailors can help users create customized products, ensuring a perfect fit and design.
- *Reward Points*: The FABIO app appreciates sustainable efforts. Users earn reward points every time they upcycle their denim jeans. These can be redeemed for exclusive discounts, eco-friendly merchandise, or charitable donations.
- *Material Marketplace*: Find all the necessary materials for upcycling in one place. FABIO app hosts a Material Marketplace, offering a curated selection of sustainable fabrics, zippers, buttons, and more to enhance your creative projects.
- *Shop Sustainable Denim*: Discover a collection of pre-loved, high-quality denim jeans available for purchase on the app. By choosing sustainable denim, users contribute to reducing fashion waste and promoting a circular economy.
- *Showcase Your Creativity*: The FABIO app provides a platform for users to showcase their upcycled denim creations to the community. Participate in monthly planned events and stand a chance to get featured on the app as the top sustainable designer.
- *Digital Sustainability*: The FABIO app is designed with scalability in mind. It can easily integrate with other e-commerce platforms, ensuring its digital sustainability and accessibility to a broader user base.
- *Eco-conscious Lifestyle*: The FABIO app welcomes all denim enthusiasts, fashion-conscious individuals and eco-warriors who strive for a sustainable lifestyle. By embracing upcycling, users become advocates for a greener and more conscious approach to fashion.
- *Social Media:* Social media influencers have emerged as key players in promoting sustainable fashion. These platforms not only facilitate networking and communication but also foster circular entrepreneurship through avenues like second-hand product trading.

Discussion and Conclusion

FABIO's circular economy model, shown in Figure 6.9, encompasses the entire jeans lifecycle, starting with sustainable farming practices that employ recycled water and bio-fertilizers. Food crop waste is collected and processed, providing farmers with additional income streams. Cotton spinning waste is channelled to the oil and cosmetic industries, optimizing resource utilization. Utilizing advanced Agraloop technology, FABIO converts cotton and food crop waste into biodegradable fabric, harmonizing fashion with sustainability. FABIO's jeans manufacturing process embraces water efficiency and waste minimization, serving as a blueprint for eco-conscious apparel production.

FABIO plays a pivotal role in upskilling labour within the fashion industry, empowering not only farmers and tailors but also designers and customers. Through active collaboration, FABIO fosters a close-knit community of fashion

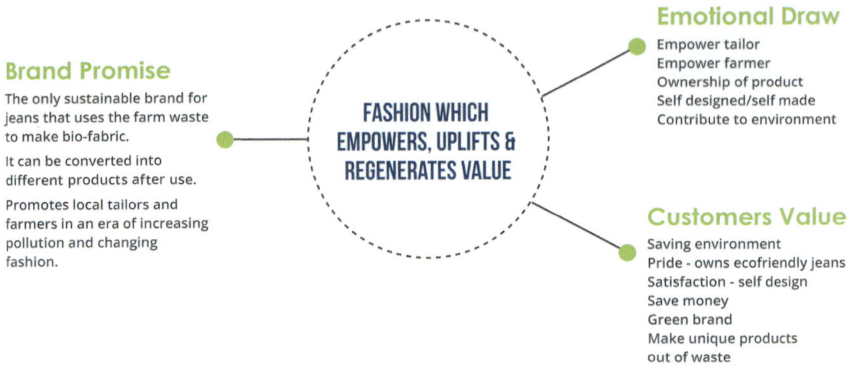

Figure 5.9 Brand promise for FABIO.

influencers, uniting individuals committed to sustainable living. This not only reduces risks associated with environmental damage and pollution but also promotes risk mitigation strategies. Moreover, FABIO's efforts extend beyond the fashion sphere, raising awareness within society about the importance of eco-conscious choices. In the realm of technology sustainability, FABIO excels in making tech solutions more efficient, exemplifying innovation in the pursuit of waste reduction and environmental conservation. In essence, FABIO's impact spans multiple dimensions, from labour empowerment and community building to risk mitigation, societal awareness and tech sustainability, all contributing to the path towards sustainable fashion. Beyond mitigating environmental burdens, FABIO cultivates a mindful and responsible lifestyle among consumers, transcending transactions to forge a meaningful connection with the environment. FABIO actively rewards environmentally responsible actions, fostering a sense of responsibility towards the planet.

Acknowledgements

We would like to express our gratitude to the Indian Institute of Information Technology, Design and Manufacturing, Jabalpur, Professor Vasant Devaji, and all the people who contributed their valuable insights and support during this project.

References

Circular Systems (2023), 'Agraloop - circular systems: Regenerative impact'. Available at: https://circularsystems.com/agraloop (accessed 7 September 2023).

Ellen MacArthur Foundation (2023a), 'Circular Economy Introduction'. Available at: https://www.ellenmacarthurfoundation.org/topics/circular-economy-introduction/overview (accessed 30 November 2023).

Ellen MacArthur Foundation (2023b), 'The Butterfly Diagram: Visualising the Circular Economy'. Available at: https://ellenmacarthurfoundation.org/circular-economy-diagram (accessed 7 September 2023).
Heshmati, A. (2017), 'A review of the circular economy and its implementation', *International Journal of Green Economics*, 11(3/4): 251–288.
Levi Strauss & Co. (2023), 'The life cycle of a Jean'. Available at: https://www.levistrauss.com/wp-content/uploads/2015/03/Full-LCA-Results-Deck-FINAL.pdf (accessed 29 November 2023).
Negrete-Cardoso, M., Rosano-Ortega, G., Álvarez-Aros, E. L., Tavera-Cortés, M. E., Vega-Lebrún, C. A. and Sánchez-Ruíz, F. J. (2022), 'Circular economy strategy and waste management: A bibliometric analysis in its contribution to sustainable development, toward a post-covid-19 era', *Environmental Science and Pollution Research*, 29(41): 61729–61746.
Papamichael, I., Chatziparaskeva, G., Pedreno, J. N., Voukkali, I., Candel, M. B. A. and Zorpas, A. A. (2022), 'Building a new mind set in tomorrow fashion development through circular strategy models in the framework of Waste Management', *Current Opinion in Green and Sustainable Chemistry*, 36: 100638.
Peter John, E. and Mishra, U. (2023), 'A sustainable three-layer circular economic model with controllable waste, emission, and wastewater from the textile and fashion industry', *Journal of Cleaner Production*, 388: 135642.
Webber, K. (2022), 'The environmental and human cost of making a pair of jeans'. Available at: https://www.ecowatch.com/environmental-cost-jeans-2544519658.html (accessed 7 September 2023).

Part II

Engineering and the Built Environment

Chapter 6

Upcycling Carbon-Fibre Reinforced Polymer Matrix Composites: An Engineering Management Perspective

Abhishek Tiwary and Dazel Empe

Introduction

Carbon Fibre Reinforced Polymer (CFRP) has rapidly become a manufacturing essential over the last decade owing to its outstanding properties, including strength-to-weight ratio, low density, high chemical and fatigue resistance and favourable anisotropic mechanical properties, offering excellent design flexibility over its conventional metal counterparts (Yang et al., 2012; van de Werken et al., 2019). Its global abundance is expected to be massive, given the forecasted increase in its production, which is expected to reach around 190,000 tonnes annually by 2050 (Zhang et al., 2020). This is estimated to lead to a cumulative total of over half a million tonnes of CFRP waste by 2050; the majority of these waste piles will be located in North America and Europe (Lefeuvre et al. 2017). Cumulative CFRP composite waste from end-of-life aircraft and wind turbine decommissioning alone is expected to reach 100,000 tonnes per year, posing a long-term sustainability challenge for their conventional reuse in the transport/energy sector via existing fibre-based recycling industries (Meng et al., 2018; Sun et al., 2015; Jensen & Skelton, 2018). Therefore, the increasing use of CFRP and the consequent increase in the volume of composite waste generated has created a level of concern about its disposal method, requiring a new paradigm of composite waste reuse through upcycling.

The EU Waste Framework outlines five waste management strategies, arranged from most to least desirable, including prevention, preparing for reuse, recycling, other recovery, and disposal (DEFRA 2011). To date, sustainability engineering has largely focused on initiatives for stakeholders' engagement in the implementation of the CFRP recycling sector in a sustainable way (Pillain et al., 2020). The heterogeneous nature of CFRP composites, however, makes it extremely difficult to determine an ideal recycling route. In addition, the cost of recycling these materials is generally high, with only a little market potential for the recyclates (Rybicka et al., 2016). Although the disposal (or landfilling) method offers the cheapest option for managing composite wastes, prevalent (and upcoming) policies and environmental legislation tend to make them the least feasible option (Rybicka et al., 2016). This opens a huge opportunity for upcycling of carbon-fibre composite materials in alternative manufacturing industries, specifically utilizing their lightweight and high-strength attributes in designing non-critical structures (Rodrigues et al., 2014).

Integrating sustainability studies in the decision-making process for reusing carbon fibres is increasingly being adopted from technical perspectives. Studies have investigated the economic and environmental performance of recycling system techniques (Pillain et al., 2020; Dahmus et al., 2008). There is rapid progress in innovative advanced engineering methods to enable high-performing materials to sustain similar or perhaps enhanced technical effectiveness of the recycled/ upcycled CFRP product (Zhu et al., 2019). The upcycling of recycled carbon fibre, CFRP waste or prepreg carbon waste has already been evaluated in terms of its mechanical properties in the production of eco-efficient cement-based materials (Akihama et al., 1986; Akbar & Liew, 2020), the reuse and upcycling of thermoset prepreg scrap in the production of composite laminates (Nilakantan & Nutt, 2017) and cement mortar reinforcement (Nguyen et al., 2016). Nevertheless, to date, there is a clear lack of a management approach to match technological advancement to translate these innovations into real-world applications. This is essential to support the increasing levels of upcycling opportunities for CFRP in the future, including infrastructure development, logistics and upcycling technological advances. To this end, appropriate legislation on the recycling and upcycling of CFRP will also have a major impact on the use of composites. Current drawbacks in financing commercial scale applications are the inadequacy of infrastructure and the lack of markets (Conroy et al., 2006).

The overall aim of this chapter is to develop an understanding of the challenges and opportunities of upcycling carbon fibre composite materials. A PESTEL-SWOT analysis was conducted to identify the key political, economic, social, technological, environmental, and legal factors that influence the opportunities and limitations to the broader adoption of CFRP upcycling. This analysis incorporates both industry trends and evidence from the relevant literature. Thereafter, the outcomes and some recommendations will be made.

Methods

PESTEL-SWOT is an integrated strategic management approach that combines the concepts of a factor-based and a strength-based analysis. It is applied to screening-level evaluation of the strengths and weaknesses of implementing innovation projects (Srdjevic et al., 2012). PESTEL analysis is an audit of the factors (internal/ external) affecting project implementation (Gray, 2016). PESTEL comprises a six-category system: Political (P), Economic (E), Social (S), Technological (T), Environmental (E) and Legal (L). Political factors represent the governmental (and intergovernmental) initiatives influencing policies and investments. Economic factors reflect segments that appear unique to the economy and how they directly influence the project. Social exhibits the demographic, socio-cultural and behavioural factors. Technological refers to the rate of technological advancements (or the lack thereof), such as research and development activities, automation, and incentives. Environmental points all the elements that concern environmental preservation. Legal specifies all the laws directly linked to the project. On the other hand, SWOT analysis is derived from the four-letter acronym representing the four key elements of strategic analysis: strengths (S), weakness (W), opportunities (O) and threats (T).

The factor extraction procedure involves the following two significant activities based on literature (Srdjevic et al., 2012): (i) a detailed SWOT/PESTEL analysis to identify the factors influencing a given system, their impacts, and their positive or negative effects on the system; and (ii) a criteria-based eliminating algorithm to filter the factors derived by SWOT/PESTEL to identify a non-inferior set of factors. As a first step, the factors were selected according to the type of influence across the PESTEL categories based on the literature survey. These PESTEL factors were then divided by SWOT analysis into two clusters. They were identified as internal and external factors according to the domain of their influence on the system – positive or negative. Subsequently, the algorithm for identifying a non-inferior set of factors employed an approach that prioritized factors that could enhance opportunities and mitigate weaknesses in the system.

The impact intensity scale applied to each factor was acquired from an expert elicitation workshop, ranging between 1–5 (1 = Very low, 2 = Low, 3 = Moderate, 4 = High, 5 = Very high). These outcomes of a PESTEL-SWOT analysis provide a list of factors useful for further analysis and for acquiring the decision measures in the multicriteria analysis (Srdjevic et al., 2012). PESTEL analysis includes the broad environmental context affecting the project and the changes that materialize in this context. SWOT analysis, on the other hand, interprets the intensity of these results for the project, identifying their strengths and weaknesses, opportunities and threats (Gray, 2016).

The scope of the integrated PESTEL-SWOT analysis in this chapter is to efficiently demonstrate both the positive and the negative aspects of developing an upcycling plan for CFRP composites. A proposed elimination algorithm filters out all given factors to identify the non-inferior factors and indicate these as candidates for the criteria set. In order to reach a solution to satisfy the goals, multicriteria instruments were employed in the forms of compensational and not-compensational methods, utility theory and other practices. The first step mainly analysed the system using the PESTEL factors. This involved assessing their impact over time (short-, medium- or long-term), by its type (positive or negative) and by dynamics (importance of the increasing, decreasing or unchanged implications) using SWOT (Gray, 2016). The second step covered the recommendations for improvement of the implementation, such as how to increase strength, lessen the weaknesses, capitalize on opportunities and mitigate identified threats (Srdjevic et al., 2012).

Results and Discussion

Trends from PESTEL-SWOT Analysis

Table 7.1 lists the key political, economic, social, technological, environmental, and legal factors influencing both the opportunities and limitations to the broader adoption of CFRP upcycling. The last two columns of the table show the SWOT outcomes, respectively indicating the directions of impact (+/—) and their intensities (scale of 1–5, showing low-to-high) across the different factors identified. The analysis reveals the following trends.

- *Political:* (i) direct role of government investments in supporting the research and infrastructure development initiatives; (ii) indirect role of climate change levy in discouraging CFRP landfill; and (iii) strong effect of lobbying from influential organizations to bring the level of acceptance of upcycled goods at par with virgin materials.
- *Economic*: The lack of an established market represents the most significant challenge yet simultaneously presents an opportunity for the value generation potential of upcycling. This is derived from the reduction in the high recovery costs of recycled CFRP and the manufacturing of virgin fibres. The incentives for avoiding costs from waste disposal levy and the contributions to corporate sustainability goals are common to those of CFRP recycling.
- ***Social:*** The potential for job creation and social responsibility (i.e. improved citizenship) from reduced waste disposal and the subsequent damage to the environment are identified as the key factors leading to public endorsement.
- *Technological:* The shortcomings of the current technology base, the dearth of skilled personnel, and the uncertainty surrounding the costs of advanced upcycling techniques represent the most significant obstacles to the advancement of upcycling. Nevertheless, the potential for technological innovation in CFRP upcycling to facilitate the emergence of new markets, extending beyond the current applications of CFRP in the manufacturing and transport sectors, is considerable.
- *Environmental*: The factors identified in this category have a significant positive impact on the broader adoption of CFRP upcycling. In particular, the offsetting of demand on virgin carbon fibres (its associated primary resource, energy, and waste disposal demands) and the alternative to hazardous chemical-based recycling represent the key environmental opportunities from upcycling.
- *Legal:* The legislation on CFRP recycling is still evolving, never mind the added complexities of including upcycling legislation, which may be only determined once prominent routes to CFRP upcycling become established. While the 'polluter pays' principle applicable to CFRP recycling can be extended to upcycling, it may not be widely accepted by the parent manufacturing companies owing to their limited end-use potential in their existing product range.

Based on the PESTEL-SWOT analysis, environmental factors are one of the most critical determinants affecting the reuse (recycling and/or upcycling) of CFRP wastes. When carbon fibres are recycled/reused to displace virgin carbon fibres, the impact is directed to virgin carbon fibre production avoidance (or minimization). These avoided impacts serve as the benchmark for defining the level of the environmental advantage (Witik et al., 2013). It is notable that most of the available literature on CFRP reuse still focuses on recycling. In some cases, the term upcycling has been blended with recycling, with no specific distinction in the approaches adopted. Only a small number of papers have exclusively covered the upcycling aspect of CFRP reuse. This demonstrates that while CFRP recycling has dominated the research landscape, with operational advancements over the past decade, the upcycling aspect is still in its infancy. Nevertheless, there appears to be an emerging shift in the role of upcycling to balance the environmental pressure from rapidly growing CFRP waste burdens (and the inadvertent release of pollution from some of the prevalent recycling techniques) with the value generation potential at a relatively lower processing cost compared to expensive recycling techniques.

Limitations to upcycling of carbon fibre and recommendations

While recycling techniques have become more prevalent, ranging from conventional to advanced processes, the concept of upcycling is still relatively new. There is currently a lack of clearly developed upcycling routes (mainly infrastructure, supply chain and logistics) linked to other material industries, as well as the insufficiency of end products and their market demand. Also, the legislation concerning recycling is mainly geared toward the process industry for recovery and utilization of composites in the parent industry (European Council, 2000) rather than their upcycling into new products. It is still unclear what type of recyclate can generate the most lucrative market for recycled CFRP waste (ibid.), let alone the new challenge of determining their upcycling potential. The uncertainties in commercial-scale applications (infrastructure and markets) for upcycled products have resulted in a lack of an adequate financing structure for entrepreneurs (Conroy et al., 2006).

A further significant challenge associated with the upcycling of CFRPs is the effective identification and separation of multiple material constituents, particularly in composites with heterogeneous structures (Rybicka et al., 2016). CFRPs heavily utilize cross-linked thermoset polymers as a matrix, which cannot be melted or remoulded. It is common practice to combine CFRPs with other types of reinforcements, such as glass and aramid fibres, core materials, metallic inserts and paint, resulting in a complex multiphase of waste types. This poses a technological challenge for the separation of CFRP from the other waste types before their further upcycling (Witik et al., 2013).

Academic research is primarily focused on different recycling techniques to recover carbon fibres extensively. However, the role of government in enforcing policies plays a vital role. Strategies such as implementing substantial levies on

Table 6.1 List of factors influencing management of carbon-fibre reinforced polymer matrix composite upcycling using PESTEL-SWOT analysis

Category	Factor	Trends and Implication	Impact type	Scale
POLITICAL	Strong R&D focus for increased government support for carbon-fibre reuse	In the US, the Carbon Fibre Recycling Act administers the aerospace and automotive industries to provide carbon fibre recycling projects and accredited $10M for R&D (U.S Senate Committee on Energy & Natural Resources, 2015). In the UK, £16M of government funding has been used to set up the National Composite Centre to support research and innovation.	+	5
	Climate Change Levy	The heavy landfill tax discourages the use of CFRP landfilling because the rates increase every year (Rybicka et al., 2016).	+	5
	Waste Treatment Policy	EU legislation limits environmental waste impacts and creates resource efficiency by promoting the use of waste as a resource (Witik et al., 2013). In the UK, improved reuse of end-of-life CFRP resources reduces landfilling activities (CIWM, 1999).	+	4
	Promoting product acceptability	Role of lobbying from influential organizations. Adequate government policy instruments encourage/promote competition to the point that upcycled goods can be sold along with those produced using virgin raw materials (Matielli Rodrigues et al., 2014).	+	5
ECONOMIC	Production cost	The necessary material and equipment costs in operations remain relatively constant across the economies of leading manufacturing sites (Das, 2011; Das et al., 2016).	+	4
	Value generation	The critical value obtained from upcycling CFRPs supports cutting back, reducing the high recovery costs of recycled CFRP, and manufacturing virgin fibres (Li et al., 2016).	+	4
	Waste disposal fee elimination	Upcycling CFRP eliminates landfilling, hence zero waste disposal fees; this transforms expensive practice into a profitable, renewable material (Pimenta and Pinho, 2011). Disposal of the CFRP through landfilling (if legal) costs approximately £0.20 per kg. Upcycling could avoid this cost, thereby creating a profitable reusable material (Sapuan, 2017).	+	5
	Setting of corporate goals	In the UK, a commercial company recycling carbon fibres has set out an ambitious plan to produce 6,500 tonnes of fibre-recovered materials before 2021 (Holmes, 2018).	+	4
	Lack of established market	Different recycling methods are made available to each industry, as recycling carbon fibres is a topical theme. However, the greatest challenge of upcycling carbon fibres is setting up a market for the composite trade (Holmes, 2018).	-	4
SOCIAL	Job creation	The local communities will have opportunities to work on manufacturing grounds during building and project life operations and maintenance.	+	5
	Public endorsement	The collaboration of different organizations will make the projects more credible, leading to more people approving CFRP upcycling.	+	4
	Social responsibility	The community benefits overall from reduced waste going to landfills through the recovery and reuse of CFRPs (Sapuan, 2017).	+	4

TECHNOLOGICAL	Infrastructure gap	Lack of reliable public facilities for identification and collection of end-of-life components for upcycling.	-	2
	Lack of ambitious end-of-life design vision	As part of retired aircraft schemes, airline operators (or their business segments) are more focused on scrapping retired aircraft at a low cost, with little or no consideration of any recycling/upcycling option (Carberry, 2008).	-	5
	Skills gap	In the past 20 years, more effort has been made in glass fibre research; carbon fibre composite recycling/upcycling is still in its infancy (Karuppannan et al., 2020).	-	3
	Cost-inefficiency of CFRP recycling techniques	The industrial application of the solvolysis process requires an expensive reactor that can withstand elevated temperatures, pressure, and corrosion (Kritzer, 2004). Methods that thermochemically break down cross-linked matrix material involve a more intricate process, thereby increasing the processing costs (Li et al., 2016).	-	4
	End-of-life consumer focus	To date, the use of carbon fibres is heavily dominated by the manufacturing and transport sectors which provides huge potential for other sectors to join and gain expertise (Composites in Manufacturing, 2016).	+	5
ENVIRONMENTAL	Primary resource demand	The demand for virgin carbon fibre often exceeds the supply capacity, thus increasing the potential for upcycling carbon fibres into innovative structures (Pimenta & Pinho, 2011).	+	5
	Energy demand	Typically, an energy requirement of 111.88 MJ/kg is needed to produce 1kg of virgin composite material. In contrast, only a tenth of that energy is required to obtain the same weight for a recycled composite material (Jensen & Skelton, 2018).	+	5
	Opportunity for environmentally friendly waste management alternatives	The cumulated carbon-fibre wastes expected by 2030 due to EoL (aircraft and wind turbines) is projected to be 100,000 tons annually, which is beyond the handling scope of the current available CFRP recycling industries (Karuppannan et al., 2020).	+	5
	Climate change mitigation	The greenhouse gas emissions produced by virgin fibre production are ten times higher than those produced by traditional metals. Recovering carbon fibre from CFRP wastes essentially helps to compensate for the production impacts (Li et al., 2016).	+	5
	Alternative to hazardous chemical-based recycling	Upcycling innovation can replace conventional chemical recycling methods, which use strong acids or solvents under different conditions and pose extreme environmental impact potentials (Karuppannan et al., 2020).	+	5
LEGAL	"Polluter pays" principle	In the EU, a 70% minimum weight requirement of non-hazardous and demolition waste shall be opted for reuse, recycling, or material recovery (European Commission, 2008).	+	5
	End of Life Vehicle Directive	Requires (as of 1 Jan 2015) 85% of manufactured cars to be reused or recycled (European Council, 2000).	+	5
	No aviation regulation	Appropriate legislation is lacking that compels aircraft owners/manufacturers to recycle their end-of-life aircraft components (Lefeuvre et al., 2017).	-	4

Note: Impact intensity scale: 1 = Very low, 2 = Low, 3 = Moderate, 4 = High, 5 = Very high composite industries for environmental degradation are directing the composite market towards the closure of the life-cycle loop of carbon fibre wastes through the implementation of sustainable recycling practices. While there is a push for the replacement of virgin fibres with recycled carbon fibres, there is also an urgent need to determine commercially viable recycling/upcycling methods in industrial-scale processes and their adequate management structures (Meng et al., 2020). The latter is needed for the establishment of CFRP upcycling on a large industrial scale to be consistently operated as an economically viable and sustainable system across all stages – recovery, transportation, separation and manufacturing of the upcycled product/s.

From a management perspective, the following are recommended to develop a viable CFRP reuse system, addressing some of the above-identified issues. A framework approach to sustainable CFRP reuse is required, with a focus on balancing recycling and upcycling techniques. This would necessitate the convening of disparate stakeholders, including suppliers, recyclers/upcyclers and researchers, with the objective of achieving a unified understanding of the current technological landscape and charting a course for future advancement in this field. In order to achieve the desired outcome, it will be necessary to ensure close collaboration between the waste suppliers, logistics and the supply chains involved. In sectors such as the aviation industry, where the majority of CFRP consumption is generated, there is a clear need for a more stringent legislative framework to ensure that private organizations are compelled to comply, regardless of the presence or absence of any incentive. In contrast to the automotive industry, there is currently no legislative directive on aircraft recycling.

Conclusion

This chapter implemented an integrated PESTEL–SWOT analysis to identify the key political, economic, social, technological, environmental and legal underpinnings of CFRP upcycling. A synthesis of the positive and negative aspects of the subject matter, in conjunction with an analysis of industry trends and a review of pertinent literature, reveals that environmental factors represent one of the most significant determinants influencing the reuse (recycling and/or upcycling) of CFRP wastes. Nevertheless, while the recycling of carbon fibre is advancing rapidly, this process comes at a significant economic and environmental cost. Recommendations based on this analysis show the need for a clearly developed upcycling route (mainly infrastructure, supply chain and logistics) linked to other process industries, as well as the creation of market demand for the upcycled end products. Furthermore, an analysis of the extant literature reveals that the prevailing sustainability evaluation framework for CFRP waste reuse is predominantly focused on recycling, with the objective of facilitating the recovery and utilization of carbon fibres in the parent industry, thereby reducing the reliance on virgin resources. In contrast, there is a paucity of literature on accounting for the attributes associated with the upcycling of CFRP waste into new products. Further investigation is required to assess the cost–benefits of different upcycling initiatives to provide a sustainable solution for the re-utilization of growing CFRP waste volumes.

Acknowledgements

This research was funded by a Global Challenge Research Fund internal grant from De Montfort University, UK, grant number 10003243.

References

Akbar, A. and Liew, K.M. (2020), 'Assessing recycling potential of carbon fiber reinforced plastic waste in production of eco-efficient cement-based materials', *Journal of Cleaner Production*, 274(November): 123001.

Akihama, S., Suenaga, T. and Banno, T. (1986), 'Mechanical properties of carbon fibre reinforced cement composites', *International Journal of Cement Composites and Lightweight Concrete*, 8(1): 21–38.

Carberry, W. (2008), 'Airplane recycling efforts benefit Boeing operators', *Boeing Aero Magazine ORT*, 4(8): 6–13.

CIWM (1999), 'Landfill Directive (1999/31/EC)'. Available at: https://www.ciwm.co.uk/ciwm/knowledge/landfill-directive.aspx.

Composites in Manufacturing (2016), 'Lightening the load – composites in manufacturing'. Available at: https://www.composites.media/lightening-the-load/.

Conroy, A., Halliwell, S. and Reynolds, T. (2006), 'Composite recycling in the construction industry', *Composites Part A: Applied Science and Manufacturing*, 37(8): 1216–1222.

Dahmus, J.B., Fredholm, S.A., Olivetti, E.A., Gregory, J.R. and Kirchain, R.E. (2008), 'Modeling the economic and environmental performance of recycling systems', *IEEE International Symposium on Electronics and the Environment*, 1–6.

Das, S. (2011), 'Life cycle assessment of carbon fiber-reinforced polymer composites', *The International Journal of Life Cycle Assessment*, 16(3): 268–282.

Das, S., Warren, J., West, D. and Schexnayder, S.M. (2016), 'Global carbon fiber composites supply chain competitiveness analysis', *Clean Energy Analysis Centre*. Available at: https://www.osti.gov/servlets/purl/1333049.

DEFRA (2011), 'Government Review of Waste Policy in England 2011'. Available at: https://assets.publishing.service.gov.uk/government/uploads/system/uploads/attachment_data/file/69401/pb13540-waste-policy-review110614.pdf.

European Commission (2008), 'European Waste Framework Directive'. Available at: http://data.europa.eu/eli/dir/2008/98/2018-07-05.

European Council (2000), 'Directive 2000/53/EC of the European Parliament and of the Council on End-of-Life Vehicles'. Available at: https://www.legislation.gov.uk/eudr/2000/53/body.

Gray, B. (2016), 'How to effectively conduct a PESTLE & SWOT analysis'. Available at: https://www.linkedin.com/pulse/how-conduct-pestle-swot-analysis-byron-gray

Holmes, M. (2018), 'Recycled carbon fiber composites become a reality', *Reinforced Plastics*, 62(3): 148–153.

Jensen, J. P. and Skelton, K. (2018), 'Wind turbine blade recycling: experiences, challenges and possibilities in a circular economy', *Renewable and Sustainable Energy Reviews*, 97(December): 165–176.

Kritzer, P. (2004), 'Corrosion in high-temperature and supercritical water and aqueous solutions: A review', *The Journal of Supercritical Fluids*, 29(1–2): 1–29.

Lefeuvre, A., Garnier, S., Jacquemin, L., Pillain, B. and Sonnemann, G. (2017), 'Anticipating in-use stocks of carbon fiber reinforced polymers and related waste flows generated by the commercial aeronautical sector until 2050', *Resources, Conservation and Recycling*, 125(October): 264–272.

Li, X., Bai, R. and McKechnie, J. (2016), 'Environmental and financial performance of mechanical recycling of carbon fibre reinforced polymers and comparison with conventional disposal routes', *Journal of Cleaner Production*, 127(July): 451–460.

Meng, F., Cui, Y., Pickering, S. and McKechnie, J. (2020), 'From aviation to aviation: Environmental and financial viability of closed-loop recycling of carbon fibre composite', *Composites Part B: Engineering*, 200(November): 108362.

Meng, F., Olivetti, E. A., Zhao, Y., Chang, J. C., Pickering, S. J. and McKechnie, J. (2018), 'Comparing life cycle energy and global warming potential of carbon fiber composite recycling technologies and waste management options', *ACS Sustainable Chemistry & Engineering*, 6(8): 9854–9865.

Nguyen, H., Carvelli, V., Fujii, T. and Okubo, K. (2016), 'Cement mortar reinforced with reclaimed carbon fibres, CFRP waste or Prepreg carbon waste', *Construction and Building Materials*, 126(November): 321–331.

Nilakantan, G. and Nutt, S. (2017), 'Reuse and upcycling of thermoset Prepreg scrap: Case study with out-of-autoclave carbon fiber/epoxy Prepreg', *Journal of Composite Materials*, 52(3): 341–360.

Pillain, B., Lefeuvre, A., Garnier, S., Cadene, A. L. and Jacquemin, L. (2020), 'Sustainability engineering assessment research for recycling composites with high value: Stakeholders' views', *Sustainable Development*, 28(1): 197–207.

Pimenta, S. and Pinho, S. T. (2011), 'Recycling carbon fibre reinforced polymers for structural applications: technology review and market outlook', *Waste Management*, 31(2): 378–392.

Puttegowda, M., Rangappa, S. M., Jawaid, M., Shivanna, P., Basavegowda, Y. and Saba, N. (2018), 'Potential of natural/synthetic hybrid composites for aerospace applications', *Sustainable Composites for Aerospace Applications*, January: 315–351.

Rodrigues, G. G. M., de Paiva, J. M. F., do Carmo, J. B. and Botaro, V. R. (2014), 'Recycling of carbon fibers inserted in composite of DGEBA epoxy matrix by thermal degradation', *Polymer Degradation and Stability*, 109(November): 50–58.

Rybicka, J., Tiwari, A. and Leeke, G. A. (2016), 'Technology readiness level assessment of composites recycling technologies', *Journal of Cleaner Production*, 112 (January): 1001–1012.

Sankar, K.G. and Kärki, T. (2020), 'A review on the recycling of waste carbon fibre/glass fibre-reinforced composites: fibre recovery, properties and life-cycle analysis', *SN Applied Sciences*, 2(3): 1–21.

Sapuan, S. M. (2017), *Composite materials: Concurrent engineering approach*, Oxford: Butterworth-Heinemann.

Srdjevic, Z., Bajcetic, R. and Srdjevic, B. (2012), 'Identifying the criteria set for multicriteria decision making based on SWOT/PESTLE analysis: A case study of reconstructing a water intake structure', *Water Resources Management*, 26(12): 3379–3393.

Sun, H., Guo, G., Memon, S. A., Xu, W., Zhang, Q., Zhu, J. H. and Xing, F. (2015), 'Recycling of carbon fibers from carbon fiber reinforced polymer using electrochemical method', *Composites Part A: Applied Science and Manufacturing*, 78 (November): 10–17.

U.S Senate Committee on Energy & Natural Resources (2015), 'S. 1432, the Carbon Fiber Recycling Act of 2015'. Available at: https://www.congress.gov/bill/114th-congress/senate-bill/1432,

Van de Werken, N., Reese, M. S., Taha, M. R. and Tehrani, M. (2019), 'Investigating the effects of fiber surface treatment and alignment on mechanical properties of recycled carbon fiber composites', *Composites Part A: Applied Science and Manufacturing*, 119 (April): 38–47.

Witik, R. A., Teuscher, R., Michaud, V., Ludwig, C. and Månson, J. A. E. (2013), 'Carbon fibre reinforced composite waste: An environmental assessment of recycling, energy recovery and landfilling', *Composites Part A: Applied Science and Manufacturing*, 49 (June): 89–99.

Yang, Y., Boom, R., Irion, B., Van Heerden, D. J., Kuiper, P. and De Wit, H. (2012), 'Recycling of composite materials', *Chemical Engineering and Processing: Process Intensification*, 51 (January): 53–68.

Zhang, J., Chevali, V. S., Wang, H. and Wang, C. H. (2020), 'Current status of carbon fibre and carbon fibre composites recycling', *Composites Part B: Engineering*, 193 (July): 108053.

Zhu, J. H., Chen, P. Y., Su, M. N., Pei, C. and Xing, F. (2019), 'Recycling of carbon fibre reinforced plastics by electrically driven heterogeneous catalytic degradation of epoxy resin', *Green Chemistry*, 21 (7): 1635–1647.

Chapter 7

Circular Economy Solutions for Solar Photovoltaic Panels: Review and Future Research for UK Scholars

Kyungeun Sung, Yi Zhong and Patrick Isherwood

Introduction

There is widespread consensus that anthropogenic climate change is one of the most pressing issues facing humanity. It is mainly caused by the burning of fossil fuels, including coal, oil and natural gas (Witzel & Seifried, 2013; Tingem & Rivington, 2009). The burning of fossil fuels provides 80 per cent of the world's energy. However, it also emits several additional greenhouse gases into the atmosphere, primarily carbon dioxide (CO_2), which is leading to an increase in the Earth's average surface temperature and has already been linked to an observed increase in the number and severity of natural disasters such as hurricanes, typhoons and rising sea levels (Kaygusuz, 2001; Quaschning, 2014). Global energy consumption is increasing significantly to meet the energy needs of a growing population, but current fossil fuel reserves, particularly oil and gas, will be depleted by the end of the twenty-first century (Kannan & Vakeesan, 2016; Quaschning, 2014). All this is driving the need to explore renewable energy sources that have the potential to meet global energy needs without harming the climate (Panwar et al., 2011; Twidell, 2021). Renewable energy has become increasingly important not only because of climate change but also, more recently, the energy crisis (Farghali et al., 2023). As a form of renewable energy, solar energy has significant advantages in terms of availability, cost-effectiveness, accessibility, energy security, capacity and efficiency, as it is abundant, non-exhaustible, free to exploit, affordable, clean and widely applicable in different scenarios (Schlamadinger et al., 1997; Nozik, 1978; Lewis, 2007). Solar energy, particularly photovoltaics (PV), had the highest annual growth rate of all renewable energy capacity in 2013 and is likely to become one of the main pillars of the global energy supply this century (Bakhiyi et al., 2014; Quaschning, 2014; Raturi, 2019).

Over the past few decades, PV has boomed in many countries and regions and is expected to continue to grow (Weckend et al., 2016). The six regions with the highest installed solar PV capacity are China, Europe (excluding Germany), North America, Japan, Germany and India (Fraunhofer Institute for Solar Energy Systems,

2023). The flourishing of PV in some of these countries is linked to financial incentives from their governments to encourage adoption (Richter, 2013). In the UK, for example, PV installations have proliferated under the influence of financial support for utilities (2010–2019), driving the transformation of the UK power sector in combination with its leading offshore wind technology (International Energy Agency, 2019). The UK government stated that a low-cost, net-zero emissions system in the UK could consist mainly of wind and solar power (HM Government, 2020). Despite the promising potential and rapid development, several stages of the PV lifecycle have a negative impact on the environment. This includes production, operation and end-of-life.

During the manufacturing phase, silicon production relies on high energy consumption (both in the extraction of silicon from silica and the conversion of metallurgical grade silicon to solar grade silicon). This energy comes mainly from the combustion of fossil fuels, contributing to greenhouse gas emissions, toxic chemical emissions and wastewater (Greenwood & Earnshaw, 2012; Ausland & Rong, 2020; Ramírez-Márquez et al., 2020). Silicon waste in the process of slicing solar-grade silicon into silicon wafers is typically over 40 per cent (Compaan, 2006). During the operational phase, the environmental impact is relatively low. However, some large-scale solar installations can have a negative impact on local habitats, such as destroying local wildlife habitats, disrupting local irrigation or livestock systems, or occasionally polluting nearby groundwater sources (Stoms et al., 2013; Aman et al., 2015). During the end-of-life phase, many damaged photovoltaic panels are sent to landfill as waste without any recycling, potentially damaging soil and groundwater (Kang et al., 2012; Bakhiyi et al., 2014). The mixed plastics used in solar panels are difficult to recycle (Xu et al., 2018). End-of-life management of PV modules is becoming increasingly important due to the increasing levels of waste. Solar panels typically have a lifespan of 20–30 years, and as the global PV market grows, the number of end-of-life PV panels will increase, creating a new form of environmental pollution (Bakhiyi et al., 2014; Weckend et al., 2016). In 2014, electronic waste amounted to 41.8 million tonnes, of which PV waste accounted for just 0.1 per cent. However, PV panel waste could exceed one-tenth of total electronic waste by 2050 (IRENA, 2016).

Recognizing both the benefits and challenges/issues of solar panels, this chapter examines available solutions to make the PV lifecycle more environmentally sustainable in line with upcycling (i.e., creative reuse, repair, refurbishment, remanufacture, recycling, etc.) and the circular economy – an alternative to the traditional, linear economy of take, make, use and dispose, making the material flow circular, slowing down the material flow/circulation, and closing the material loops (Stahel, 2016; Ellen MacArthur Foundation, 2016). This review focuses on addressing end-of-life management as a critical challenge.

Context: UK and the Circular Economy Potential

PV manufacturing in the United Kingdom is relatively small, with most processes taking place outside the country. The industry is largely dependent on imports to supply modules, particularly from China (GOVUK, 2013). The UK government has provided financial support for the development of PV modules as well as module installation (International Energy Agency, 2019; GOVUK, 2013). Based on historical cumulative installed PV capacity, it is estimated that the amount of end-of-life PV panels in the UK will be between 30,000 and 200,000 tonnes by 2030 and 1 to 1.2 million tonnes by 2050 (IRENA, 2016). For end-of-life PV panels, the UK was one of the first countries to accept the European Union's categorisation of waste PV components as E-waste under Waste Electrical and Electronic Equipment (WEEE) legislation and to require all key data to be provided for imported panels (European Union, 2012; IRENA, 2016). The UK's WEEE Directive includes photovoltaic panels as a separate category and encourages the recovery, reuse and recycling of waste products and components (UK Government, 2021). PV producers/suppliers in the UK are usually responsible for funding the collection of B2C (Business to Consumers) PV panels and for funding the collection and recycling of B2B (Business to Business) PV panels (UK Government, 2023; IRENA, 2016). Although e-waste must be treated in accordance with the principles of the UK WEEE legislation, in practice, end-of-life PV modules are more likely to be treated in other central European countries due to the lack of available high-quality treatment facilities in the UK (IRENA, 2016).

Although the growth of solar waste may pose environmental problems in the future, it could also present several opportunities for value creation from a circular economy perspective. The circular economy is a restorative and regenerative system in which resource use, waste, emissions and energy leakage are minimized by slowing down, closing and narrowing material and energy loops. This can be achieved through durable design, maintenance, repair, reuse, remanufacture, refurbishment and recycling (Geissdoerfer et al., 2017). The circular economy emphasizes cradle-to-cradle upcycling processes, such as end-of-life PV panel components becoming resources without loss of quality and being used to make new products (Contreras-Lisperguer et al., 2017). In particular, recycling end-of-life solar panels can effectively conserve natural resources and reduce production costs (Xu et al., 2018). Using components recovered from decommissioned solar panels can save two-thirds of the energy used in the original production process (Aman et al., 2015). There are several materials from end-of-life solar panels that could bring huge economic benefits by 2030. For example, silicon (estimated at around 30,000 tonnes) has a potential value of around \$50 million in the secondary materials market and could be used to make at least 45 million new panels. Silver (approximately 90 tonnes) has a potential value of \$180 million and could be used to make up to 50 million new panels. Aluminium (approximately 75,000 tonnes) and copper (about 7,000 tonnes) have a combined potential value of up to \$140 million, and glass (approximately 960,000 tonnes) has a potential value of over \$28 million (IRENA, 2016).

The majority of PV modules that are currently recycled are typically those that were defective at the time of manufacture and/or have been damaged during transport, installation, or operation (Tao & Yu, 2015). There are currently not many end-of-life PV panels in the UK, and the PV panel recycling market is likely to remain small in the immediate future (IRENA, 2016). However, there will be a dramatic increase in the number of end-of-life PV modules in the coming decade. Industrial-scale facilities for the treatment of end-of-life PV modules are scarce at present, and the current ideal recycling process remains technically and economically challenging (Tao & Yu, 2015). Therefore, the potential value of end-of-life PV modules and the potential environmental risks highlight the need for immediate attention to the development of PV module recycling facilities and processes in the UK. This will be in line with the UK's national strategic goals since the UK Government released a Clean Growth Strategy in 2017 with a series of policies and proposals to meet their emission reduction goals (UK Government, 2017) and launched a Circular Economy Package in 2020, identifying steps for the reduction of waste, reaffirming the ambition to recycle 65 per cent of municipal waste by 2035, and aiming to move towards a more circular economy (UK Government, 2020).

Recycling of PV Panels

A typical solar energy system consists of solar panels and an inverter for grid connection (Xu et al., 2018). Crystalline silicon (c-Si) accounted for over 90 per cent of global PV production in 2022 (Fraunhofer Institute for Solar Energy Systems, 2023). A typical c-Si solar panel comprises an aluminium frame, silicon cells, a junction box, a glass front sheet, a glass or polymer back sheet, and a laminate encapsulant, usually EVA (ethylene vinyl acetate) (Wang et al., 2022; Deng et al., 2019). A c-Si solar panel has six materials with recycling potential: (i) glass (69–75% weight percentage in a module) which has relatively low economic value; (ii) aluminium (10–20%) with medium value; (iii) copper (4.4–7%) with high value; (iv) silicon (2–3%) with high value; and (v) silver (0.006–0.08%) with high value (Wang et al., 2022). While glass and aluminium make up a large proportion of recycled PV panel materials, copper, silicon and silver have a higher economic value. Reclaiming silicon wafers is deemed to be the most significant challenge among these materials, given that their manufacturing cost comprises 65 per cent of producing silicon cells, which, in turn, contribute to 60 per cent of the total PV module manufacturing cost (Calì et al., 2022; Kang et al., 2012). Recovering silicon from recycled solar panels could save a third of the energy and cost compared to direct silicon manufacturing (Choi & Fthenakis, 2010). Despite only accounting for a small percentage of PV panels, silver has a high economic value and limited future supplies, highlighting the significance of silver recovery and recycling (Kuczyńska-Łażewska et al., 2018; Sverdrup et al., 2014). The recovery of silicon and silver can be regarded as the two high-priority areas for recycling to attain a significant level of economic sustainability.

The technologies and techniques required for the recycling of PV modules have largely been developed at laboratory scale but have not been scaled up until recently (Chowdhury et al., 2020; Tao & Yu, 2015). Australia appears to have some emerging solar waste management and recycling companies, such as PV Industries in Sydney and Pan Pacific Recycling in Crestmead. The recycling process for a c-Si PV module typically starts with the removal of the aluminium frame and junction box (Olson et al., 2013). The frameless module undergoes decontamination and subsequently goes through the processes of delamination and recovery of silicon and other valuable components (Isherwood, 2022; Wang et al., 2022). Module delamination de-bonds EVA from the module, making it easier to recycle silicon, glass and other valuable materials. This is typically achieved through a combination of mechanical, thermal and chemical processes (Isherwood, 2022; Wang et al., 2022; Tao & Yu, 2015). Mechanical delamination could be used, for example, in abrasive machining under cryogenic conditions and electrostatic separation. However, based on this process, the recycled silicon is of low purity and cannot be reused in the production of new silicon wafers (Zhang et al., 2013). For thermal delamination, a two-step heating process could be used to burn the EVA and obtain a direct reuse glass under a well-controlled temperature (Wang et al., 2012; Fiandra et al., 2019). Many of the thermal processes require more than two stages to eradicate the EVA and require an additional process to recycle the silicon wafers, resulting in high costs, energy consumption and harmful gas emissions (Tao & Yu, 2015). For chemical delamination, typically nitric acid, solvents or solvents coupled with ultrasonic irradiation are used to dissolve EVA, including the use of organic solvents (Kang et al., 2012). Some of these chemical processes either require a prolonged treatment time or need to be combined with additional mechanical and thermal treatments to eradicate the EVA, resulting in negative environmental impact due to the use of large quantities of chemical reagents (most of which are toxic) and the waste fluids produced. Most of the aforementioned delamination processes are still in the research phase due to significant environmental concerns, such as high energy costs, toxic gas emissions, and chemical waste disposal (Tao & Yu, 2015). In addition, EVA swells quickly during thermal or chemical processing and can lead to degradation and breakage of the silicon cells (Yu & Yang, 2013). Further research is needed to improve the effectiveness and economy of processes or technologies for removing EVA and recovering undamaged silicon wafers and glass. Although the technologies required to remove EVA are a critical part of the development of methods (Xu et al., 2018), some other research directions have also been suggested, such as the development of environmentally friendly alternatives to current polymer materials or the development of technologies to recycle the polymer materials (EVA or back-sheet) (Isherwood, 2022). More recent studies have explored alternative options, such as using TPO (Thermoplastic Polyolefin) instead of EVA (Schnatmann et al., 2022) or laser irradiation method instead of any thermal or chemical processes (Li et al., 2022).

After module delamination, the silicon is separated from the other bulk materials but is still aggregated with the metal current collection layers, anti-reflection coating materials and dopants. Consequently, further treatments are required to purify the silicon to achieve high-grade material (Xu et al., 2018). Where silicon can be removed as complete wafers, etching and leaching have proven to be effective purification methods, for example, alkali-acid leaching baths to remove copper, lead, silver, aluminium and dopants from the silicon (Eshraghi et al., 2020; Tao & Yu, 2015). Several etching processes with varying solutions were tested, which demonstrated that the recovered silicon cells could have a good efficiency of 13–15% without an anti-reflective coating (Klugmann-Radziemska et al., 2010; Klugmann-Radziemska & Ostrowski, 2010). Some wafer recovery and reprocessing techniques can produce solar-grade silicon, but the expense of recovery makes it less economically viable (Calì et al., 2022). As PV technology has developed, silicon wafer thickness has decreased; preventing them from breaking during recycling may become increasingly challenging (Heath et al., 2020; Domínguez & Geyer, 2019; Wang et al., 2022). In terms of recovery of other valuable materials, a leaching solution could be used to dissolve silver, lead, copper and tin. These materials can then be recovered using techniques such as electrowinning (Huang et al., 2017). Overall, the recovery of materials from end-of-life PV modules has potential environmental and economic benefits. However, the potential negative environmental impacts of the chemicals required and the potentially expensive and complex processes involved require further consideration (Mahmoudi et al., 2020; Ardente et al., 2019; Padoan et al., 2019).

In addition to recycling the valuable materials from end-of-life PV modules, redesigning the end-of-life PV modules to make them easier to recycle or longer lasting could also be an alternative way to achieve a circular economy. Redesigning PV modules to make their components easier to recycle or reuse has potential benefits in both environmental and economic terms when compared to the complex material recycling process of end-of-life PV modules (Tao & Yu, 2015). For instance, a novel double encapsulation module with an optical coupler has been proposed for relatively quickly recyclable silicon cells from end-of-life PV modules and reducing the electrical performance loss compared to the standard encapsulation scheme (Li et al., 2011). Optimal and convenient materials, geometries and tolerances for the couplings between layers and components in PV panels have been identified to find the best current, efficient and effective solutions for recycling end-of-life panels and to improve panel durability (Calì et al., 2022).

Other Circular Economy Solutions for Solar PV Panels

Recycling is the most researched of the circular economy solutions (or upcycling options) for PV panels to date, as it is generally considered to be the most viable option when the panels reach the end of their life. This is largely a function of their design, which can make repair, reuse, refurbishment, etc. difficult (Weckend et al., 2016; Gobbi, 2011; Komoto et al., 2018). However, focusing on recycling as the only viable option is short-sighted and will not lead to a reliably circular PV industry. This would go against the basic idea of the circular economy, which is that products should be used for as long as possible, with recycling being the last resort (El-Khawad et al., 2022). As circular economy solutions in general, there are 10Rs of sustainability, including 6Rs which embeds 3Rs (Bag et al., 2021; del Río et al., 2021). 3Rs are (i) reduce, (ii) reuse, and (iii) recycle while 6Rs have additional

(iv) refuse, (v) rethink and (vi) repair (Sakthivel et al., 2021; Potting et al., 2017). The four additions to 10Rs are (vii) refurbish, (viii) remanufacture, (ix) repurpose, and (x) recover. In 3Rs: (i) reduce is optimizing and decreasing the consumption of natural resources in manufacturing, and (ii) reuse is utilizing discarded PV modules that are still in working condition. In 6Rs: (iv) refuse means rejecting harmful processes or proposing sustainable alternatives; (v) rethink is restructuring and optimizing the design and life-cycle planning of PV modules, and (vi) repair is maintaining defective PV modules to their best possible functionality. In 10Rs: (vii) refurbish means restoring older PV modules to the highest technical specification possible; (viii) remanufacture means utilizing discarded parts in manufacturing new PV modules and technologies; (xi) repurpose means utilizing discarded parts in manufacturing new products for a different purpose (other than PV panels); and (x) recover means optimizing processes to recover as many materials and energy as possible (Rabaia et al., 2022). In terms of end-of-life management, typically, out of the 10Rs, the repair and reuse strategies are the most favourable, followed by refurbishing and remanufacturing, then recycling and, lastly, incineration and/or landfill (Mathur et al., 2019).

Although repair is the most favourable end-of-life management strategy for the circular economy, traditional solar PV modules are often not readily repairable. The design is a plug-and-play solution, and the end users are rarely able to repair the panel in the event of damage (Majdi et al., 2021; El-Fayome et al., 2023). As the fundamental issue is the embedded irreparability, new designs and innovations are essential for future solar panels, such as development of a modular design with the ability to replace any parts in case of damage without the need to replace the entire panel (Majdi et al., 2021), or novel robust encapsulation design which nonetheless provides easy and direct access to individual components for repairs, upgrades, or modifications (El-Fayome et al., 2023). For the current panels, a sensing and repair system could be utilized to detect the state of each cell in the module and reconfigure connections between them (e.g., a memristor-based sensing and repair system as a modular solution that could be potentially embedded inside a solar panel, leading to a self-healing device) (Gnoli et al., 2021). To detect PV panels with faults requiring repair or replacement, thermal imaging processing software based on artificial intelligence technology could be used for preventative maintenance (Constantin et al., 2023). The recently suggested non-contact soldering method could be a viable option for fixing damage to solar panel interconnections. This method does not compromise the performance of the solar cell while generating a solder connection without cracks during extended exposure times (Goodarzi et al., 2023).

Reuse represents another most advantageous practice for the circular economy. Many decommissioned solar panels remain functional but are significantly less efficient than modern technologies, limiting reuse purposes due to the considerable increase in efficiency in recent years (El-Khawad et al., 2022). However, the experimental case study of reusing abandoned solar panels as a power supply for mobile phone charging stations and signal boosters in an off-grid agricultural village in Indonesia (Pandyaswargo et al., 2022) shows the potential for reuse of

PV modules in similar situations. In the case of panel reuse, they must be certified with, for example, an electrical compliance certificate or an equivalent certification (providing information about the capacity and age of the second-hand panels) for the safe and efficient operation of used panels (Majewski & Dias, 2023). The same applies to the refurbished or remanufactured panels as well. Currently, there is no second-hand panel recertification standard, which is the major problem. Repair, reuse refurbishment and remanufacturing of waste modules encounter limitations in discovering a local market, decreased efficiency, shortened lifetime, and inconsistency with current quality and safety standards (Tao et al., 2020). In particular, the acceptance of reused, refurbished, or remanufactured (RRR) products is a crucial issue, as consumers may expect consistent power output and high-efficiency rates. It is also known that consumers tend to be sceptical towards RRR products in general (Wallner et al., 2020). As RRR solar panels are still uncommon on the market, there remains a lack of understanding regarding consumer acceptance and purchase behaviour (El-Khawad et al., 2022). It would also be crucial for suppliers of RRR products to furnish sufficient information and data substantiating any claims regarding the efficacy and safety of such products to overcome the market challenges. This is where a standard would help, too, by providing a minimum standardized guarantee of quality and lifetime.

Despite the availability of some promising options, the PV industry continues to employ a take-make-use-dispose approach (Salim et al., 2019; et al., 2021). This is due to issues in the PV value chain, such as no design consideration for re-x processes, no product service system, geographically limited take-back plans, a lack of communication throughout the supply chain stages, a lack of plans for voluntary product circulation by customers, informal recyclers, limited dedicated PV dismantling facilities, delay of effective end-of-life management planning, and landfill costs currently cheaper than dismantling (Rabaia et al., 2022). Therefore, industry stakeholders need to consider: (i) designing future solar panels to ensure easy and affordable re-x processes, (ii) exploring the option of a product as a service (El-Khawad et al., 2022), (iii) providing a product stewardship scheme or extended producer responsibility; (iv) banning landfills for solar panels (e.g., Australia, EU, UK) (Majewski et al., 2021); and (v) exploring the life cycle symbiosis (extension of industrial symbiosis to identify waste streams with value as potential raw material/feedstock, through the development of collaborative business/economic interactions among a diverse set of organizations) (Mathur et al., 2020). To address the challenges faced in the PV industry's value chain, effective collaboration between parties is essential for achieving circularity (e.g., producers, importers, dealers, system operators, waste treatment companies, and end users) (Sica et al., 2018).

Discussion and Conclusion

Considerable research and development activities have focused on material savings, panel efficiencies, forward-flow supply chain perspectives, and recycling, mainly from a technological point of view, with little regard for circular design, circular business models and PV re-x processes other than recycling (Franco & Groesser, 2021; Sica et al., 2018; Rabaia et al., 2022). While acknowledging the significance of circular design in general, designing new PV modules for the circular economy may not be the most beneficial course of action for UK scholars for further research due to its relatively small PV manufacturing base. Circular business model development, on the other hand, is a highly relevant and crucial area that the UK's future researchers should focus on, particularly with regard to various re-x processes (as part of upcycling). To maximize usability and minimize the misuse of PV waste, thorough and in-depth research should be conducted to establish the criteria for classifying the different conditions of end-of-life PV modules (Rabaia et al., 2022).

Although there has been extensive research on recycling PV modules globally, the UK currently relies on treatment in other central European countries due to the lack of high-quality treatment facilities in the UK. Therefore, future research should prioritize understanding the current status quo of the PV value chain in the UK, benchmarking the best practices in central Europe for potential replication in the UK, and developing ready-to-deploy pathways with reproducible processes and supply chain systems with a usable classification and recertification system for PV module waste. The process of benchmarking, potential replication of the best practices and development of ready-to-deploy pathways should involve the UK's PV industry stakeholders as key partners for co-design and co-development of the UK's recycling system. For the establishment of the recycling system/facilities, one should prioritize the recovery of silicon and silver (for economic sustainability) and focus on scaling up the best practices from the lab scale while improving the effectiveness and economy of processes/technologies. Testing and proving economic viability and minimizing the negative environmental impacts of the technology/facility are critical to the process.

Along with the development of the recycling system in the UK, the potential of other re-x processes could be explored. For example, what repair options are viable for large-scale or domestic installations in the UK? What kinds of reuse of end-of-life PV modules could be economically feasible in the UK? How could we certify reused/refurbished/remanufactured PV modules in the UK? What would be the recertification standard, process and criteria? What should be the minimum threshold for ensuring the safe and efficient operation of used panels? What potential options exist for repurposing PV waste in the UK? How can PV waste be reused for purposes beyond the PV value chain? (e.g., for building/construction materials or street furniture). How can we address the limitations of RRR PV panels, such as market discovery, reduced efficiency, shortened lifespan and potential consumer reluctance to accept RRR products? Despite the limited review of the existing literature, this chapter recognizes the crucial advancements in circular economy solutions for solar PV panels, offering a review contextualized within the UK framework and outlining potential areas for future research by UK scholars.

Acknowledgements

This research was funded by the AHRC Research Networking grant for the International Upcycling Research Network, grant number AH/W007134/1. The De Montfort University's (DMU) Net Zero and Climate Action fund also partially supported this research. The work was supported by the generous Research Award from DMU. We thank Nicky Harding, Stuart Lawson, Kate Cheyne, Kelley Wilder, Siobhan Keenan, Christine White, Jill Cowley, Deborah Cartmell, Mike Kagioglou and Katie Normington for supporting this research at DMU.

References

Aman, M.M., Solangi, K.H., Hossain, M.S., Badarudin, A., Jasmon, G.B., Mokhlis, H., Bakar, A. and Kazi, S.N. (2015), 'A review of Safety, Health and Environmental (SHE) issues of solar energy system', *Renewable and Sustainable En-ergy Reviews*, 41: 1190–1204.

Ardente, F., Latunussa, C.E. and Blengini, G.A. (2019), 'Resource efficient recovery of critical and precious metals from waste silicon PV panel recycling', *Waste Management*, 91: 156–167.

Ausland, G. and Rong, H.M. (2020), 'Silicon production with high energy efficiency and low CO_2 footprint', in B. Andresen, H.R. Elkem, M. Tangstad, H. Tveit, I.G. Page (eds), *Silicon for the Chemical and Solar Industry XV*, 13–22, Trondheim: The Norwegian University of Science and Technology.

Bag, S., Gupta, S. and Kumar, S. (2021), 'Industry 4.0 adoption and 10R advance manufacturing capabilities for sustainable development', *International Journal of Production Economics*, 231: 107844.

Bakhiyi, B., Labrèche, F. and Zayed, J. (2014), 'The photovoltaic industry on the path to a sustainable future—Environmental and occupational health issues', *Environment International*, 73: 224–234.

Calì, M., Hajji, B., Nitto, G. and Acri, A. (2022), 'The Design Value for Recycling End-of-Life Photovoltaic Panels', *Applied Sciences*, 12(18): 9092.

Choi, J. and Fthenakis, V. (2010), 'Design and optimization of photovoltaics recycling infrastructure', *Environmental science & technology*, 44(22): 8678–8683.

Chowdhury, M.S., Rahman, K.S., Chowdhury, T., Nuthammachot, N., Techato, K., Akhtaruzzaman, M., Tiong, S.K., Sopian, K. and Amin, N. (2020), 'An overview of solar photovoltaic panels' end-of-life material recycling', *Energy Strategy Reviews*, 27: 100431.

Compaan, A.D. (2006) 'Photovoltaics: Clean power for the 21st century', *Solar Energy Materials and Solar Cells*, 90(15): 2170–2180.

Constantin, A., Iosif, G., Chihaia, R., Marin, D., Abu Shehadeh, G.U., Karahan, M., Gerikoglu, B. and Stavrev, S. (2023) 'Importance of preventive maintenance in solar energy systems and fault detection for solar panels based on thermal images', *Electrotehnica, Electronica, Automatica*, 71(1): 1–12.

Contreras-Lisperguer, R., Muñoz-Cerón, E., Aguilera, J. and de La Casa, J. (2017), 'Cradle-to-cradle approach in the life cycle of silicon solar photovoltaic panels', *Journal of Cleaner Production*, 168: 51–59.

del Río, P., Kiefer, C.P., Carrillo-Hermosilla, J., Könnölä, T., del Río, P., Kiefer, C.P., Carrillo-Hermosilla, J. and Könnölä, T. (2021), 'The micro-level approach to the circular economy', in P. del Río, C.P. Kiefer, J. Carrillo-Hermosilla, T. Könnölä (eds), *The Circular Economy: Economic, Managerial and Policy Implications*, 73–87, Cham: Springer.

Deng, R., Chang, N.L., Ouyang, Z. and Chong, C.M. (2019), 'A techno-economic review of silicon photovoltaic module recycling', *Renewable and Sustainable Energy Reviews*, 109: 532–550.

Domínguez, A. and Geyer, R. (2019), 'Photovoltaic waste assessment of major photovoltaic installations in the United States of America', *Renewable Energy*, 133: 1188–1200.

El-Fayome, E., Zekry, A.H., Abdelhamed, M.A., EL-Shazly, A., Abouelatta, M., Shaker, A. and Saeed, A. (2023), 'Proposal for Repairable Silicon Solar Panels: Proof of Concept', *Energies*, 16(18): 6492.

El-Khawad, L., Bartkowiak, D. and Kümmerer, K. (2022), 'Improving the end-of-life management of solar panels in Germany', *Renewable and Sustainable Energy Reviews*, 168: 112678.

Ellen MacArthur Foundation (2016), 'Circular Economy'. Available at: http://www.ellenmacarthurfoundation.org

Eshraghi, N., Berardo, L., Schrijnemakers, A., Delaval, V., Shaibani, M., Majumder, M., Cloots, R., Vertruyen, B., Boschini, F. and Mahmoud, A. (2020), 'Recovery of nano-structured silicon from end-of-life photovoltaic wafers with value-added applications in lithium-ion battery', *ACS Sustainable Chemistry & Engineering*, 8(15): 5868–5879.

European Union (2012), 'Directive 2012/19/EU of the European Parliament and of the Council on waste electrical and electronic equipment (WEEE)'. Available at: /https://eur-lex.europa.eu/LexUriServ/LexUriServ.do?uri=OJ:L:2012:197:0038:0071:en:PDF

Farghali, M., Osman, A.I., Mohamed, I.M., Chen, Z., Chen, L., Ihara, I., Yap, P. and Rooney, D.W. (2023), 'Strategies to save energy in the context of the energy crisis: a review', *Environmental Chemistry Letters*, 21(4): 2003–2039.

Fiandra, V., Sannino, L., Andreozzi, C., Corcelli, F. and Graditi, G. (2019), 'Silicon photovoltaic modules at end-of-life: Removal of polymeric layers and separation of materials', *Waste Management*, 87: 97–107.

Franco, M.A. and Groesser, S.N. (2021), 'A systematic literature review of the solar photovoltaic value chain for a circular economy', *Sustainability*, 13(17): 9615.

Fraunhofer Institute for Solar Energy Systems (2023), 'Photovoltaics report'. Available at: https://www.ise.fraunhofer.de/en/publications/studies/photovoltaics-report.html

Geissdoerfer, M., Savaget, P., Bocken, N.M. and Hultink, E.J. (2017), 'The Circular Economy – A new sustainability paradigm?', *Journal of Cleaner Production*, 143: 757–768.

Gnoli, L., Riente, F., Ottavi, M. and Vacca, M. (2021), 'A memristor-based sensing and repair system for photovoltaic modules', *Microelectronics Reliability*, 117: 114026.

Gobbi, C. (2011), 'Designing the reverse supply chain: the impact of the product residual value', *International Journal of Physical Distribution & Logistics Management*, 41(8): 768–796.

Goodarzi, D.M., Lauri, J., Putaala, J., Nousiainen, O. and Fabritius, T. (2023), 'Eddy current soldering of solar cell ribbons under a layer of glass', *Solar Energy Materials and Solar Cells*, 259: 112427.

GOV.UK (2013), 'UK Solar PV Strategy - UK Solar PV Strategy Part 1: Roadmap to a Brighter Future', *Department of Energy & Climate Change and The Rt Hon Gregory*

Barker. Available at: https://www.gov.uk/government/publications/uk-solar-pv-strategy-part-1-roadmap-to-a-brighter-future

Greenwood, N.N. and Earnshaw, A. (2012), *Chemistry of the Elements*. London: Elsevier.

Heath, G.A., Silverman, T.J., Kempe, M., Deceglie, M., Ravikumar, D., Remo, T., Cui, H., Sinha, P., Libby, C. and Shaw, S. (2020), 'Research and development priorities for silicon photovoltaic module recycling to support a circular economy', *Nature Energy*, 5(7): 502–510.

HM Government (2020), 'Energy White Paper: Powering Our Net Zero Future'. Available at: https://assets.publishing.service.gov.uk/media/5fdc61e2d3bf7f3a3bdc8cbf/201216_BEIS_EWP_Command_Paper_Accessible.pdf

Huang, W., Shin, W.J., Wang, L., Sun, W. and Tao, M. (2017), 'Strategy and technology to recycle wafer-silicon solar modules', *Solar Energy*, 144: 22–31.

International Energy Agency (2019), 'Energy policies of IEA countries'. Available at: https://iea.blob.core.windows.net/assets/298930c2-4e7c-436e-9ad0-2fb8f1cce2c6/Energy_Policies_of_IEA_Countries_United_Kingdom_2019_Review.pdf

IRENA (2016), 'End-of-life management: Solar Photovoltaic Panels'. Available at: https://www.irena.org/publications/2016/Jun/End-of-life-management-Solar-Photovoltaic-Panels

Isherwood, P.J. (2022), 'Reshaping the module: the path to comprehensive photovoltaic panel recycling', *Sustainability*, 14(3): 1676.

Kang, S., Yoo, S., Lee, J., Boo, B. and Ryu, H. (2012), 'Experimental investigations for recycling of silicon and glass from waste photovoltaic modules', *Renewable Energy*, 47: 152–159.

Kannan, N. and Vakeesan, D. (2016), 'Solar energy for future world: A review', *Renewable and sustainable energy reviews*, 62: 1092–1105.

Kaygusuz, K. (2001), 'Renewable energy: Power for a sustainable future', *Energy Exploration & Exploitation*, 19(6): 603–626.

Klugmann-Radziemska, E. and Ostrowski, P. (2010), 'Chemical treatment of crystalline silicon solar cells as a method of recovering pure silicon from photovoltaic modules', *Renewable Energy*, 35(8): 1751–1759.

Klugmann-Radziemska, E., Ostrowski, P., Drabczyk, K., Panek, P. and Szkodo, M. (2010), 'Experimental validation of crystalline silicon solar cells recycling by thermal and chemical methods', *Solar Energy Materials and Solar Cells*, 94(12): 2275–2282.

Komoto, K., Lee, J., Zhang, J., Ravikumar, D., Sinha, P., Wade, A. and Heath, G.A. (2018), 'End-of-life management of photovoltaic panels: Trends in PV module recycling technologies', *Photovoltaic Power Systems Programme*. Available at: https://www.osti.gov/servlets/purl/1561523.

Kuczyńska-Łażewska, A., Klugmann-Radziemska, E., Sobczak, Z. and Klimczuk, T. (2018), 'Recovery of silver metallization from damaged silicon cells', *Solar Energy Materials and Solar Cells*, 176: 190–195.

Lewis, N.S. (2007) 'Toward cost-effective solar energy use', *Science*, 315(5813): 798–801.

Li, H., Luo, Y., Ballif, C. and Perret-Aebi, L. (2011), 'Re-use of c-Si solar cells from failed PV modules', *Proceedings of the 26th EU-PVSEC Conference*, Hamburg.

Li, X., Liu, H., You, J., Diao, H., Zhao, L. and Wang, W. (2022), 'Back EVA recycling from c-Si photovoltaic module without damaging solar cell via laser irradiation followed by mechanical peeling', *Waste Management*, 137: 312–318.

Mahmoudi, S., Huda, N. and Behnia, M. (2020), 'Environmental impacts and economic feasibility of end-of-life photovoltaic panels in Australia: A comprehensive assessment', *Journal of Cleaner Production*, 260: 120996.

Mahmoudi, S., Huda, N. and Behnia, M. (2021), 'Multi-levels of photovoltaic waste management: A holistic frame-work', *Journal of Cleaner Production*, 294: 126252.

Majdi, A., Alqahtani, M.D., Almakytah, A. and Saleem, M. (2021), 'Fundamental study related to the development of modular solar panel for improved durability and repairability', *IET Renewable Power Generation*, 15(7): 1382–1396.

Majewski, P., Al-shammari, W., Dudley, M., Jit, J., Lee, S., Myoung-Kug, K. and Sung-Jim, K. (2021), 'Recycling of solar PV panels-product stewardship and regulatory approaches', *Energy Policy*, 149: 112062.

Mathur, N., Singh, S. and Sutherland, J.W. (2020), 'Promoting a circular economy in the solar photovoltaic industry using life cycle symbiosis', *Resources, Conservation and Recycling*, 155: 104649.

Mathur, N., Deng, S., Singh, S., Yih, Y. and Sutherland, J.W. (2019), 'Evaluating the environmental benefits of implementing industrial symbiosis to used electric vehicle batteries', *Procedia CIRP*, 80: 661–666.

Nozik, A.J. (1978), 'Photoelectrochemistry: applications to solar energy conversion', *Annual Review of Physical Chemistry*, 29(1): 189–222.

Olson, C.L., Geerligs, L.J., Goris, M., Bennett, I.J. and Clyncke, J. (2013), 'Current and future priorities for mass and material in silicon PV module recycling', *EU-PVSEC*, Paris.

Padoan, F.C., Altimari, P. and Pagnanelli, F. (2019), 'Recycling of end-of-life photovoltaic panels: A chemical prospective on process development', *Solar Energy*, 177: 746–761.

Pandyaswargo, A.H., Wibowo, A.D. and Onoda, H. (2022), 'Reusing solar panels to improve access to information and communication in an off-grid village: A financial feasibility assessment', *Energy Reports*, 8: 857–865.

Panwar, N.L., Kaushik, S.C. and Kothari, S. (2011), 'Role of renewable energy sources in environmental protection: A review', *Renewable and sustainable energy reviews*, 15(3): 1513–1524.

Potting, J., Hekkert, M.P., Worrell, E. and Hanemaaijer, A. (2017), 'Circular economy: measuring innovation in the product chain', *PBL Netherlands Environmental Assessment Agency*. Available at: chrome-extension://efaidnbmnnnibpcajpcglclefindmkaj/https://www.pbl.nl/uploads/default/downloads/pbl-2016-circular-economy-measuring-innovation-in-product-chains-2544.pdf.

Quaschning, V. (2014), *Understanding renewable energy systems*. Abingdon: Routledge.

Rabaia, M.K.H., Semeraro, C. and Olabi, A. (2022), 'Recent progress towards photovoltaics' circular economy', *Journal of Cleaner Production*, 373: 133864.

Ramírez-Márquez, C., Villicaña-García, E., Cansino-Loeza, B., Segovia-Hernández, J.G. and Ponce-Ortega, J.M. (2020), 'Inherent occupational health hazards in the production of solar grade silicon', *Process Safety and Environmental Protection*, 142: 285–294.

Raturi, A.K. (2019), 'Renewables 2019 global status report', *REN21*. Available at: https://www.ren21.net/gsr-2019/,

Richter, L. (2013), 'Social effects in the diffusion of solar photovoltaic technology in the UK', *University of Cambridge*, Available at: https://www.repository.cam.ac.uk/items/3926264e-ec0c-4907-a390-955d6adb846f.

Sakthivel, T.S., Seal, S. and Ramakrishna, S. (2021), 'Sustainable Materials and The Circular Economy: The circular economy and sustainability are multifaceted concepts and thus should be pursued through collective advancements across various scientific fields.', *Advanced Materials & Processes*, 179(7): 28–32.

Salim, H.K., Stewart, R.A., Sahin, O. and Dudley, M. (2019), 'Drivers, barriers and enablers to end-of-life management of solar photovoltaic and battery energy storage systems: A systematic literature review', *Journal of Cleaner Production*, 211: 537–554.

Schlamadinger, B., Apps, M., Bohlin, F., Gustavsson, L., Jungmeier, G., Marland, G., Pingoud, K. and Savolainen, I. (1997), 'Towards a standard methodology for greenhouse gas balances of bioenergy systems in comparison with fos-sil energy systems', *Biomass and Bioenergy*, 13(6): 359–375.

Schnatmann, A.K., Schoden, F. and Schwenzfeier-Hellkamp, E. (2022), 'Sustainable PV Module Design—Review of State-of-the-Art Encapsulation Methods', *Sustainability*, 14(16): 9971.

Sica, D., Malandrino, O., Supino, S., Testa, M. and Lucchetti, M.C. (2018), 'Management of end-of-life photovoltaic panels as a step towards a circular economy', *Renewable and Sustainable Energy Reviews*, 82: 2934–2945.

Stahel, W.R. (2016), 'The circular economy', *Nature*, 531(7595): 435–438.

Stoms, D.M., Dashiell, S.L. and Davis, F.W. (2013), 'Siting solar energy development to minimize biological impacts', *Renewable Energy*, 57: 289–298.

Sverdrup, H., Koca, D. and Ragnarsdottir, K.V. (2014), 'Investigating the sustainability of the global silver supply, reserves, stocks in society and market price using different approaches', *Resources, Conservation and Recycling*, 83: 121–140.

Tao, J. and Yu, S. (2015), 'Review on feasible recycling pathways and technologies of solar photovoltaic modules', *Solar Energy Materials and Solar Cells*, 141: 108–124.

Tao, M., Fthenakis, V., Ebin, B., Steenari, B., Butler, E., Sinha, P., Corkish, R., Wambach, K. and Simon, E.S. (2020), 'Major challenges and opportunities in silicon solar module recycling', *Progress in Photovoltaics: Research and Applications*, 28(10): 1077–1088.

Tingem, M. and Rivington, M. (2009), 'Adaptation for crop agriculture to climate change in Cameroon: turning on the heat', *Mitigation and Adaptation Strategies for Global Change*, 14: 153–168.

Twidell, J. (2021), *Renewable energy resources*. Abingdon: Routledge.

UK Government (2017), 'Clean Growth Strategy'. Available at: https://www.gov.uk/government/publications/clean-growth-strategy

UK Government (2020), 'Circular Economy Package policy statement'. Available at: https://www.gov.uk/government/publications/circular-economy-package-policy-statement

UK Government (2021), 'Regulations: Waste Electrical and Electronic Equipment (WEEE)'. Available at: https://www.gov.uk/guidance/regulations-waste-electrical-and-electronic-equipment#full-publication-update-history

UK Government (2023), 'Electrical and electronic equipment (EEE): producer responsibilities'. Available at: https://www.gov.uk/guidance/electrical-and-electronic-equipment-eee-producer-responsibility.

Wallner, T.S., Magnier, L. and Mugge, R. (2020), 'An exploration of the value of timeless design styles for the consumer acceptance of refurbished products', *Sustainability*, 12(3): 1213.

Wang, T., Hsiao, J. and Du, C. (2012), 'Recycling of materials from silicon base solar cell module', *38th IEEE Photovoltaic Specialists Conference*, 2355–2358.

Wang, X., Tian, X., Chen, X., Ren, L. and Geng, C. (2022), 'A review of end-of-life crystalline silicon solar photovoltaic panel recycling technology', *Solar Energy Materials and Solar Cells*, 248: 111976.

Weckend, S., Wade, A. and Heath, G.A. (2016), 'End of life management: solar photovoltaic panels', *IRENA*. Available at: https://www.osti.gov/servlets/purl/1561525

Witzel, W. and Seifried, D. (2013), *Renewable Energy–The Facts*. Abingdon: Routledge.

Xu, Y., Li, J., Tan, Q., Peters, A.L. and Yang, C. (2018), 'Global status of recycling waste solar panels: A review', *Waste Management*, 75: 450–458.

Yu, S.R. and Yang, Q.Y. (2013), 'Improvement on recycling process and life cycle assessment of photovoltaic panel', *Proceedings of the EcoDesign 2013 International Symposium*, 4–6.

Zhang, J., Lv, F., Ma, L.Y. and Yang, L.J. (2013), 'The status and trends of crystalline silicon PV module recycling treatment methods in Europe and China', *Advanced Materials Research*, 724: 200–204.

Chapter 8

INTEGRATING UPCYCLING INTO URBAN INFRASTRUCTURE AND BUILT ENVIRONMENT DESIGN: A STRATEGIC APPROACH TO WASTE REDUCTION AND RESOURCE CONSERVATION

Nnyaladzi Tema and Benjamin Bolaane

Introduction

Upcycling waste has emerged as a transformative strategy within sustainable urban development, offering a practical solution to the dual challenges of waste reduction and resource conservation. Unlike conventional recycling, which often results in downcycling – where materials lose their quality and value over time, upcycling retains or enhances the material's properties, thereby extending its lifecycle and improving overall sustainability. Within the context of urban infrastructure and the built environment, upcycling plays a pivotal role in minimizing construction and demolition waste while fostering innovation in material reuse. By embracing upcycling principles, urban planners, architects and policymakers can advance the circular economy models that prioritize waste-to-resource conversion, ultimately contributing to enhanced environmental resilience and long-term sustainability.

Urban infrastructure is the cornerstone of the urban built environment, serving as the foundation for economic, social, and environmental activities. It encompasses a wide array of physical structures and systems, such as transportation networks, utilities (water supply, waste-water management, sanitation, and electricity), communication systems and public spaces, all of which are vital for the functioning of urban areas (Bigio & Dahia, 2004). The economic potential of infrastructure is vast; it not only supports economic activities but also drives them, thereby enabling the efficient movement of goods, services, and people. Access to reliable utilities such as electricity and water is fundamental for industrial development, facilitating the effective operation of businesses and industries (Arimah, 2017). Furthermore, well-designed urban infrastructure is integral to improving residents' quality of life and promoting social inclusion. Access to essential services such as clean water, sanitation, healthcare and education, contributes to better health outcomes and enhances overall well-being (Asikhia & Uyoyoghene, 2011). Public spaces, parks, and recreational facilities also play a vital role in fostering community engagement and social cohesion.

Sustainable urban infrastructure is increasingly recognized for its ability to mitigate environmental impacts and promote resilience to climate change. For example, efficient public transportation systems can alleviate energy consumption, traffic congestion and air pollution by encouraging the use of mass transit over private vehicles. Green infrastructure, such as rain gardens and permeable pavements, can manage stormwater runoff and reduce the urban heat island effect. Additionally, equitable access to infrastructure services, particularly in underserved areas, is an effective tool for reducing social disparities and promoting inclusivity. Investments in these areas, including improved transportation, healthcare facilities, and affordable housing, can address socioeconomic inequalities and enhance social mobility, contributing to the creation of more resilient and equitable urban environments.

In the context of developing countries, urban infrastructure plays a critical role in supporting sustainable development goals and enhancing the quality of life for rapidly urbanizing populations. Over the past few decades, many developing nations have witnessed substantial migration to urban centres in search of better opportunities, creating a pressing need for robust infrastructure to accommodate this growth (Green, 2014). Among the most urgent priorities are investment in water and sanitation infrastructure which are critical for public health and environmental sustainability, particularly in areas with relatively arid climates. Similarly, improvements in transportation infrastructure, including road networks and public transit systems, are necessary for reducing congestion, improving economic productivity, and fostering social mobility (Lem, 2002).

In many developing nations, there is also a growing interest in renewable energy sources, particularly solar power, to diversify energy supplies and reduce dependence on fossil fuels (Saad et al., 2024). Furthermore, as the world becomes increasingly interconnected, the importance of digital infrastructure – such as broadband internet and telecommunications networks cannot be overstated. These investments are essential for promoting innovation, education and access to information, and for ensuring economic competitiveness and social inclusion (Xu et al., 2019).

Given these considerations, prioritizing investments in urban infrastructure is essential for advancing sustainable urban development. By enhancing access to and improving the quality of urban infrastructure, developing countries can unlock economic growth, improve social well-being and build resilient communities capable of withstanding future challenges. Through the integration of upcycling practices, urban infrastructure can evolve into a more resource-efficient, sustainable, and innovative model, contributing to a more circular and regenerative urban system.

Urban Planning, Design and Infrastructure Provision

Urban planning and design plays a pivotal role in determining how effectively upcycling can be integrated into infrastructure development. Through urban layouts, streetscapes and public spaces, cities can embed upcycled materials to create resilient, resource-efficient and aesthetically appealing environments. For instance, pathways constructed from reclaimed bricks or seating areas designed with repurposed timber exemplify how upcycled materials can merge functionality with sustainability. Incorporating upcycling principles into urban planning not only minimizes landfill dependency but also strengthens the circular economy model, enabling cities to thrive while reducing their environmental footprint.

Key planning tools, such as zoning, compact urban design and recycling infrastructure, provide a robust framework for embedding upcycling into sustainable urban development. Smart zoning practices, including mixed-use zoning, support sustainable communities by fostering vibrant neighbourhoods where residential, commercial, and recreational spaces coexist (Angelopoulos et al., 2019). This reduces transportation needs, minimizes greenhouse gas emissions, and promotes local economies. By aligning zoning policies with upcycling principles, cities can incentivize the reuse of materials within mixed-use developments, further reducing the environmental impact of construction.

Compact urban design, characterized by denser developments and pedestrian-friendly environments, offers significant opportunities to integrate upcycled materials. This approach optimizes land use while minimizing resource consumption, waste generation, and the ecological footprint of urban areas (Bibri et al., 2020; McDonald et al., 2023). Urban designs that prioritize upcycled elements – such as reclaimed materials in construction and streetscape enhancements – contribute to resource conservation while enhancing the resilience of urban ecosystems. Additionally, compact cities require less infrastructure per capita, which further reduces the demand for raw materials and energy, aligning with circular economy principles (Fertner & Grobe, 2016).

The integration of waste separation and recycling infrastructure into urban design is critical for promoting a culture of resource conservation and enabling upcycling at that scale. By strategically incorporating designated spaces for recycling and material recovery within urban layouts, cities can facilitate the collection, separation, and repurposing of materials, thereby closing the loop on waste management (Sanchez-Garcia et al., 2024). Upcycling further enhances these systems by transforming discarded materials into higher-value products, driving innovation, creating employment opportunities, and strengthening community resilience to waste management challenges.

Urban planning, design and infrastructure provision are therefore indispensable in advancing sustainable cities that embrace upcycling as a cornerstone of resource conservation. By integrating upcycled materials into zoning policies, compact urban designs, and recycling facilities, urban areas can significantly reduce waste, conserve resources and foster long-term environmental sustainability (Yigitcanlar & Kamruzzaman, 2019; Addas, 2023).

Sustainable Construction Practices

The construction sector, as a major contributor to global waste production, presents a significant opportunity for integrating upcycling into sustainable construction practices. By adopting upcycling strategies, the industry can reduce environmental degradation, enhance material efficiency, and promote resource conservation. Upcycled materials such as reclaimed wood, recycled concrete, repurposed steel and salvaged glass are increasingly being incorporated into construction projects, demonstrating their potential to maintain structural integrity while offering aesthetic and functional benefits. Advancements in material processing technologies, such as bio-bricks made from agricultural waste or composite panels formed from upcycled plastics, further illustrate the transformative potential of upcycling in driving sustainable innovation within the built environment.

Green building standards provide a critical framework for embedding upcycling into sustainable construction practices. Standards such as Leadership in Energy and Environmental Design (LEED) and the Building Research Establishment Environmental Assessment Method (BREEAM) promote energy efficiency, waste reduction, and the use of environmentally friendly materials (Azhgaliyeva & Rahut, 2023; Hussain & Sheikh, 2023). By adhering to these standards, construction projects can prioritize the use of upcycled materials alongside recycled and renewable resources, optimizing resource use while minimizing waste generation. This integration not only reduces the environmental impact of construction activities but also improves the health, comfort and productivity of building occupants by ensuring better indoor air quality and efficient resource utilization.

The use of recycled and locally sourced materials further strengthens sustainable construction practices, offering both environmental and economic advantages. Incorporating upcycled materials such as reclaimed wood and recycled concrete diverts significant volumes of waste from landfills, reducing the demand for virgin resources and the energy required for raw material extraction and processing (Bolden et al., 2013; Lawrence, 2015). Additionally, sourcing materials locally minimizes transportation-related emissions, supports regional economies and fosters community resilience. When construction projects incorporate upcycling as a standard practice, they create a closed-loop system where materials are reused in innovative ways, contributing to the circular economy and reducing the construction sector's carbon footprint.

Overall, sustainable construction practices that integrate upcycling can transform the built environment into a resource-efficient, low-waste and resilient system. By embracing green building standards, prioritizing recycled and locally sourced materials and leveraging technological advancements in material processing, the construction industry can significantly reduce its environmental footprint while supporting economic and environmental sustainability.

Efficient Infrastructure Systems

Integrating upcycling into urban infrastructure systems enhances efficiency by reducing reliance on virgin raw materials, minimizing waste, and promoting sustainability. Infrastructure elements such as roads, bridges, and utility networks increasingly incorporate upcycled materials to improve durability and environmental performance. For instance, recycled plastic-infused asphalt not only offers enhanced resilience and longevity compared to conventional paving materials but also reduces maintenance costs and diverts plastic waste from landfills. Similarly, repurposed construction debris can be used in road sub-base layers, decreasing the demand for newly quarried aggregates. These practices align with sustainable urbanism objectives, fostering infrastructure systems that prioritize resource conservation, waste reduction and environmental stewardship.

Efficient infrastructure systems are critical for promoting sustainability in urban environments by optimizing resource use and mitigating environmental degradation (Oloruntobi et al., 2023). Integrating upcycled materials into infrastructure systems complements other sustainable strategies, such as smart water management and energy-efficient technologies, to build resilient and resource-conscious cities.

Incorporating upcycled materials into water infrastructure, such as rainwater harvesting systems with reclaimed components or efficient irrigation systems built from repurposed materials, reduces both environmental impact and reliance on virgin resources. These systems enable cities to harness rainwater for landscaping, irrigation and non-potable uses, thereby decreasing the strain on municipal water supplies and promoting sustainable water use (Ertop et al., 2023; Waseem et al., 2023). Advanced irrigation methods, like drip irrigation with upcycled components, further minimize water wastage and optimize plant water uptake, ensuring sustainable water management in urban landscapes. Such practices enhance urban water resilience, mitigate drought risks and preserve scarce water resources for future generations.

Upcycling also supports the development of energy-efficient urban systems. For example, street lighting infrastructure can incorporate upcycled metals or other salvaged materials in light pole manufacturing, while utilizing energy-efficient LED technologies to reduce electricity consumption and greenhouse gas emissions (Pardo-Bosch et al., 2022). LED lighting is not only more durable and cost-effective but also complements the environmental benefits of upcycling by reducing the demand for new materials. Similarly, transportation systems can benefit from upcycling, such as repurposing old materials to construct bicycle lanes or pedestrian walkways, fostering active transportation modes like cycling and walking. These energy-efficient solutions, combined with investments in electric vehicle (EV) infrastructure and public transit systems, reduce reliance on fossil fuels, lower air pollution, and minimize urban carbon footprints (Jelti et al., 2023).

By incorporating upcycling principles into smart water management and energy-efficient urban systems, cities can build infrastructure that is resource-efficient, resilient and environmentally sustainable. Upcycling not only addresses waste management challenges but also drives innovation in infrastructure design, contributing to long-term urban sustainability while fostering a circular economy.

Green Spaces and Biodiversity

The integration of upcycled materials into the development of green spaces and biodiversity-enhancing initiatives offers a sustainable approach to urban ecological design. Urban parks, community gardens, and recreational areas can benefit from upcycled elements, such as benches constructed from reclaimed wood, playground equipment made from repurposed rubber, and planters fabricated from salvaged metals. Vertical gardens and rooftop greenery projects also leverage upcycling by incorporating repurposed scaffolding structures and growing mediums derived from organic waste. These innovative applications reduce waste streams while fostering urban ecological resilience, enhancing biodiversity and addressing environmental challenges such as air pollution and the urban heat island effect.

Urban greenery plays a pivotal role in promoting biodiversity and enhancing ecological resilience within cities (Colding & Barthel, 2013). By integrating green spaces such as parks, gardens, and urban forests into the built environment, cities create essential habitats and food sources for a wide array of plant and animal species. Upcycled materials can enhance the functionality and aesthetic appeal of these spaces, such as using recycled materials to create garden beds or wildlife shelters. These green spaces also provide critical ecosystem services, including mitigating the urban heat island effect through shade provision and temperature regulation, and improving air quality by filtering pollutants (Chun & Guldmann, 2018). Additionally, incorporating upcycled materials into urban greenery supports resource conservation and sustainable urbanism by reducing dependency on virgin resources.

Ecological corridors are vital for connecting fragmented habitats and facilitating species movement within urban environments. Greenways, riparian buffers, and wildlife corridors can incorporate upcycled materials, such as bridges made from recycled steel or pathways constructed with reclaimed concrete. These interconnected networks of natural habitats serve as 'wildlife highways', enabling species to migrate, disperse and adapt to changing environmental conditions (Rudnick et al., 2012). The use of upcycled materials in designing these corridors not only reduces waste but also contributes to sustainable construction practices, ensuring that conservation efforts align with urban sustainability goals. By enhancing ecological connectivity, cities can preserve native species diversity, support genetic exchange among wildlife populations and promote resilience to urbanization and climate change (Perry & Cox, 2024).

Strategic planning and community engagement are integral to successfully integrating upcycling into green infrastructure and biodiversity projects. Urban green spaces and ecological corridors, built with upcycled materials, can inspire community participation by demonstrating innovative approaches to sustainability. Residents can engage in initiatives such as community gardening using upcycled planters or contributing waste materials for local ecological projects. These efforts not only foster a sense of environmental stewardship but also enhance urban residents' connection to nature, improving well-being and promoting social cohesion (Douglas et al., 2017).

Incorporating upcycling into urban greenery and ecological design, cities can create vibrant, resource-efficient landscapes that enhance biodiversity, mitigate environmental impacts, and improve the quality of life for urban residents. These strategies represent a forward-thinking approach to sustainable urban development, ensuring the conservation of natural ecosystems and the sustainable use of resources for future generations.

Waste Management and the Circular Economy

Upcycling plays a pivotal role in advancing circular economy principles, which focus on minimizing waste by extending product lifecycles and repurposing materials for continued economic use. As a cornerstone of sustainable urban development, upcycling can be integrated into waste management systems to divert materials from landfills and reintegrate them into value chains. Through the adoption of upcycling practices, cities can enhance resource efficiency, reduce environmental impacts and foster the development of a resilient urban economy.

Integrated waste management systems are critical for the effective handling of urban waste streams, facilitating the reduction of landfill reliance and the promotion of resource conservation. These systems should prioritize upcycling alongside traditional strategies such as recycling, composting, and waste-to-energy technologies. Upcycling-specific infrastructure, such as municipal reclamation centers for construction materials or industrial symbiosis networks, can enable the transformation of waste into high-value products. For example, construction debris can be upcycled into aggregate for road sub-base layers or artistic installations, while industrial byproducts, such as sawdust or ash, can be repurposed into innovative materials like composite panels or bricks. The inclusion of upcycling facilities within waste management systems aligns with circular economy objectives, offering cities the opportunity to reduce waste streams and extend the utility of discarded materials (Bandh et al., 2024).

Recycling facilities play a central role in separating and processing materials such as glass, metals, and plastics, while composting facilities convert organic waste into nutrient-rich soil amendments that support agricultural productivity (Grant et al., 2020). Upcycling complements these practices by focusing on enhancing the value and utility of waste materials without significant degradation in quality. For example, plastics recovered through recycling can be upcycled into durable components for urban furniture, such as benches or planters, thereby reducing reliance on virgin materials. Similarly, glass bottles can be repurposed into decorative tiles or construction materials, further embedding circular economy principles into urban design. By integrating upcycling into the broader recycling infrastructure, cities can diversify their approaches to waste transformation, creating innovative solutions that prioritize both resource efficiency and economic value creation.

Policy-driven initiatives are essential for mainstreaming upcycling within urban waste management systems. Municipal governments can incentivize businesses, industries and households to adopt upcycling practices through financial rewards, tax rebates, or grants for integrating upcycled materials into production processes. Successful examples include programmes that support the reclamation of materials from construction sites and industrial waste-sharing platforms that facilitate inter-sectoral reuse. These initiatives align with circular economy principles by encouraging waste minimization and material repurposing while promoting economic innovation and resilience (Appendino et al., 2021). Furthermore, regulatory frameworks can mandate the inclusion of upcycled materials in urban infrastructure projects, ensuring waste reduction is embedded into the planning and construction processes.

A shift towards circular economy models in urban environments requires restructured systems that minimize waste generation and emphasize product lifecycle extension. Encouraging producers to design products for longevity, reuse and recyclability is critical to achieving this objective (Yazdani & Lakzian, 2023). For instance, product designs that allow for easy disassembly can facilitate material recovery and upcycling at the end of their lifecycle. Additionally, fostering public awareness about upcycling and circular economy practices can promote sustainable consumption patterns among residents. Community-driven initiatives, such as workshops on creating upcycled products or local markets for second-hand and upcycled goods, can further reinforce these principles while fostering environmental stewardship.

Integrating upcycling and circular economy principles into urban systems can help cities achieve significant environmental and economic benefits. These practices reduce reliance on resource extraction, lower energy consumption, and mitigate pollution associated with production and waste-disposal processes (Morseletto, 2020). At the same time, they stimulate innovation, create employment opportunities in upcycling industries, and promote sustainable production and consumption patterns. By embedding upcycling into waste management and urban planning frameworks, cities can transition towards a resource-efficient, low-carbon future that prioritizes environmental sustainability and economic resilience. Through the integration of upcycling into waste management systems, urban areas can address the dual challenges of waste reduction and resource conservation while contributing to the development of a sustainable circular economy. These strategies not only enhance urban environmental performance but also provide the foundation for innovative and resilient urban economies.

Community Mobilizations and Participation

Community mobilization and participation are integral to the successful adoption of upcycling initiatives and broader sustainability efforts in urban environments. By actively involving residents in upcycling practices, cities can foster a culture of collective responsibility for waste reduction, resource conservation and environmental stewardship. Public engagement in upcycling can be promoted through grassroots initiatives, educational programmes, participatory planning, and economic incentives that empower individuals and communities to contribute meaningfully to sustainability objectives.

Grassroots initiatives, such as material exchange networks, urban maker spaces, and do-it-yourself (DIY) upcycling workshops, serve as effective platforms for engaging communities in sustainability practices. These initiatives empower residents to repurpose waste materials into valuable products, thereby reducing waste and extending the lifecycle of resources. Urban maker spaces, for instance, provide communal areas equipped with tools and resources for crafting upcycled goods, encouraging innovation and creativity among participants. Similarly, material exchange networks facilitate the redistribution of reusable items, connecting individuals and businesses with surplus materials to those who can repurpose them. Such programmes promote local ownership of sustainability efforts, fostering a sense of community and shared responsibility for environmental outcomes (Kubanza, 2020).

Educational programmes and public awareness campaigns are essential for equipping communities with the knowledge and skills they need to embrace upcycling practices. Workshops, community events, and campaigns focused on upcycling techniques, sustainable living and waste reduction can inspire residents to adopt environmentally friendly behaviours (Borawska, 2017). For example, hosting workshops on transforming discarded materials into furniture, art installations or garden equipment can demonstrate the practicality and benefits of upcycling while encouraging community engagement. Public art installations created from upcycled materials can serve both as educational tools and visual symbols of sustainability, inspiring residents to rethink their relationship with waste. By fostering awareness and promoting upcycling as a viable solution, cities can cultivate long-term behavioural changes that align with sustainability goals.

Participatory planning processes that involve residents in the development of upcycling initiatives can enhance the inclusivity and effectiveness of sustainability efforts. Municipal governments can engage communities in co-designing public spaces that incorporate upcycled elements, such as benches, planters, or playground equipment made from repurposed materials. These projects not only reduce waste but also strengthen social cohesion by encouraging collaboration among diverse stakeholders. Additionally, fostering partnerships between local businesses, schools, and residents can create synergies that amplify the impact of upcycling efforts. For instance, businesses can donate surplus materials to schools for educational upcycling projects or collaborate with community groups to develop upcycled products for local markets. Such partnerships integrate upcycling into the social and economic fabric of urban areas, enhancing their viability and sustainability.

Providing economic incentives is a powerful strategy for motivating businesses and individuals to adopt upcycling and other sustainable practices. Municipal governments can offer financial incentives, such as grants, subsidies, or tax rebates, for businesses and community organizations that integrate upcycling into their operations. For example, subsidies for establishing upcycling workshops or tax breaks for businesses using upcycled materials in production can stimulate innovation and investment in sustainable practices (White et al., 2019). Similarly, residents can be incentivized through rebates for purchasing upcycled goods or participating in community upcycling programmes. By aligning economic incentives with environmental objectives, cities can accelerate the adoption of upcycling, promote sustainable consumption, and drive systemic change in urban waste management (Goetz, 2010).

Promoting active community participation in upcycling can cultivate a culture of sustainability and environmental responsibility in cities. Public engagement in upcycling practices fosters a shared commitment to waste reduction and resource conservation while stimulating creativity and innovation in addressing environmental challenges. Through grassroots initiatives, educational campaigns, participatory planning, and economic incentives, urban areas can empower communities to take an active role in shaping a sustainable future. As residents become more engaged in upcycling and sustainability efforts, cities can transition toward more resource-efficient and resilient urban development models, ensuring long-term environmental and social benefits.

Demand Management

Effective demand management is essential to promoting the long-term viability and integration of upcycled materials into urban development. By aligning consumption patterns and urban sustainability frameworks with the circular economy principles, demand management strategies can play a critical role in reducing waste, conserving resources, and fostering environmental resilience. Policies encouraging eco-conscious procurement, sustainable design mandates, and responsible consumption practices can significantly shape market demand for upcycled products, ensuring their incorporation into mainstream development practices.

Demand-side management strategies that prioritize upcycled materials in urban development can stimulate their adoption in construction, infrastructure, and design projects. Municipal governments can introduce financial incentives, such as tax breaks or grants, for developers and businesses that utilize upcycled materials in construction and manufacturing processes. For example, upcycled concrete, reclaimed wood, and repurposed metals can be promoted as viable alternatives to conventional building materials, reducing both material waste and environmental impacts. Regulatory frameworks requiring the use of upcycled components in public procurement and infrastructure projects can further strengthen market demand. Additionally, integrating upcycled materials into urban greening initiatives – such as public parks, rooftop gardens, and playgrounds – can showcase their functionality and aesthetic appeal, encouraging their broader adoption (Rajani et al., 2022).

Consumer awareness of the environmental and economic benefits of upcycling is essential to fostering demand for sustainably produced goods. Public education campaigns, workshops, and awareness programmes can inform residents about the potential of upcycled products to contribute to waste reduction and resource conservation. Highlighting the benefits of upcycled goods – such as lower environmental footprints, cost savings, and contributions to local economies – can influence consumer preferences and drive demand for sustainable alternatives. For instance, municipal initiatives showcasing successful upcycling projects or providing subsidies for purchasing upcycled goods can encourage residents to adopt eco-conscious consumption patterns. Furthermore, collaborative campaigns with local businesses, schools, and non-governmental organizations can amplify the visibility and appeal of upcycled solutions, fostering a culture of sustainability at the community level (Price & Joseph, 2000).

Demand management strategies that promote upcycling can also be integrated into broader sustainability initiatives, such as urban mobility and water resource management. In the context of transportation, demand management approaches – such as Urban Vehicle Access Regulations (UVARs), parking management policies, and on-demand shared mobility services – can reduce congestion and emissions while encouraging the use of eco-friendly transport modes (Ogunkunbi & Meszaros, 2023). Incorporating upcycled materials into transportation infrastructure, such as bike lanes, bus shelters and pedestrian pathways, can further enhance the sustainability of urban mobility projects. Similarly, water demand management strategies, such as greywater and rainwater harvesting systems, can integrate upcycled materials like repurposed barrels or salvaged piping, reducing costs while promoting sustainable water use (Arfanuzzaman & Rahman, 2017).

Regulatory and policy-driven approaches are critical to embedding upcycling within demand management frameworks. Sustainable design mandates can require architects and developers to incorporate a minimum percentage of upcycled materials in new projects, fostering systemic change in construction practices. Urban planning policies can further promote the reuse and upcycling of materials through zoning incentives, green building certifications, and the establishment of material recovery facilities. For example, city-level initiatives supporting industrial symbiosis – whereby the waste streams of one industry become raw materials for another – can promote the large-scale integration of upcycled resources into production cycles. By creating an enabling environment for upcycling, policymakers can transform waste management systems and urban development practices into drivers of circular economy principles (Mittal & Biswas, 2019).

Demand management therefore plays a crucial role in aligning urban development practices with the circular economy goals by fostering market demand for upcycled materials, promoting responsible consumption, and optimizing resource use. Through targeted incentives, public awareness efforts, and regulatory frameworks, cities can ensure the integration of upcycling into sustainable urban development. By advancing demand-side strategies that prioritize resource conservation and waste reduction, urban areas can transition toward resilient and environmentally sustainable futures, benefiting both present and future generations.

Conclusion

The integration of upcycling into urban infrastructure and the built environment offers a promising and innovative solution to the challenges of waste management and resource conservation in contemporary urban settings. By transforming waste into high-value materials, cities can significantly reduce their environmental impact, enhance sustainability and build resilient urban systems. The adoption of upcycling practices in urban planning, construction and infrastructure design highlights its potential to reframe waste as an asset, contributing to the creation of regenerative built environments.

To advance sustainable urban development, it is essential to embed upcycling principles into policy frameworks, industry standards and societal behaviors. This will require collaborative efforts among governments, communities and stakeholders, ensuring that upcycling becomes an integral part of urban design and planning. Smart zoning, compact design, waste separation and recycling facilities can further promote resource efficiency, waste reduction and environmental stewardship, which are critical components of resilient urban systems.

Sustainable construction practices, particularly those that embrace green building standards and prioritize the use of recycled materials, are key to reducing environmental footprints and fostering resource efficiency. By supporting investments in sustainable urban infrastructure and aligning industry practices with circular economy principles, cities can create environments that are not only environmentally responsible but also economically prosperous and socially equitable. Ultimately, the successful integration of upcycling in urban development hinges on a collective commitment to sustainability. Through continued collaboration and innovation, cities can build a future where resource conservation, waste reduction, and environmental protection are at the heart of urban growth, ensuring a better quality of life for present and future generations.

References

Addas, A. (2023), 'The concept of smart cities: A sustainability aspect for future urban development based on different cities', *Frontiers in Environmental Science*, 11: 1241593.

Angelopoulos, S., Boymal, J. and De Silva, A. (2020), 'Economic benefits of 20-minute neighbourhoods', *Department of Environment, Water, Land and Planning*. Available at: https://researchrepository.rmit.edu.au/esploro/outputs/report/Economic-benefits-of-20-minute-neighbourhoods/9921861250801341.

Apipuchayakul, N. and Vassanadumrongdee, S. (2020), 'Factors affecting the consumption of energy-efficient lighting products: Exploring purchase behaviors of Thai consumers', *Sustainability*, 12(12): 4887.

Appendino, F., Roux, C., Saadé, M., and Peuportier, B. (2021), 'The circular economy in urban projects: A case study analysis of current practices and tools.' *Transactions of the Association of European Schools of Planning*, 5, 71–83.

Arfanuzzaman, M. and Rahman, A.A. (2017), 'Sustainable water demand management in the face of rapid urbanization and ground water depletion for social–ecological resilience building', *Global Ecology and Conservation*, 10: 9–22.

Arimah, B., (2017), 'Infrastructure as a catalyst for the prosperity of African Cities', *Procedia Engineering*, 198: 245–266.

Asikhia, M.O. and Uyoyoghene, O.T. (2011), 'Urban infrastructure and quality of life: A case study of Warri Metropolis', *The Nigerian Journal of Research and Production*, 19(2): 1–10.

Azhgaliyeva, D. and Rahut, D.B. (2022), 'Promoting green buildings: Barriers, solutions, and policies', *Asian Development Bank Institute (ADBI)*. Available at: https://www.econstor.eu/bitstream/10419/267764/1/1813290679.pdf.

Bandh, S.A., Malla, F.A., Wani, S.A. and Hoang, A.T. (2024), 'Waste Management and Circular Economy', in *Waste Management in the Circular Economy* (pp. 1–17). Cham: Springer

Barros, M. V., Salvador, R., do Prado, G. F., de Francisco, A. C. and Piekarski, C. M. (2021), 'Circular economy as a driver to sustainable businesses', *Cleaner Environmental Systems*, 2: 100006.

Berardi, U. (2012), 'Sustainability Assessment in the Construction Sector: Rating Systems and Rated Buildings', *Sustainable Development*, 20: 411–424.

Bibri, S.E., Krogstie, J. and Kärrholm, M. (2020), 'Compact city planning and development: Emerging practices and strategies for achieving the goals of sustainability', *Developments in the Built Environment*, 4: 100021.

Bigio, A.G. and Dahiya, B. (2004), *Urban environment and infrastructure: Toward liveable cities*, Washington: World Bank Publications.

Bolden, J., Abu-Lebdeh, T. and Fini, E. (2013), 'Utilization of recycled and waste materials in various construction applications', *American Journal of Environmental Science*, 19(1): 14–24.

Borawska, A. (2017), 'The role of public awareness campaigns in sustainable development', *Economic and Environmental Studies*, 17(4): 865–877.

Carmona, M. (2018), 'Place value: place quality and its impact on health, social, economic and environmental outcomes', *Journal of Urban Design*, 24(1): 1–48.

Chun, B. and Guldmann, J. (2018), 'Impact of greening on the urban heat island: Seasonal variations and mitigation strategies', *Computers, Environment and Urban Systems*, 71: 165–176.

Colding, J. and Barthel, S. (2013), 'The potential of 'Urban Green Commons' in the resilience building of cities', *Ecological Economics*, 86: 156–166.

Douglas, O., Lennon, M. and Scott, M., (2017), 'Green space benefits for health and well-being: a life-course approach for urban planning, design and management', *Cities*, 66: 53–62.

El-Nwsany, R.I., Maarouf, I. and Abd el-Aal, W. (2019), 'Water management as a vital factor for a sustainable school', *Alexandria Engineering Journal*, 58(1): 303–313.

Ertop, H., Kocięcka, J., Atilgan, A., Liberacki, D., Niemiec, M. and Rolbiecki, R. (2023), 'The Importance of Rainwater Harvesting and Its Usage Possibilities: Antalya Example (Turkey)', *Water*, 15(12): 2194.

Eskandaripour, H. and Boldsaikhan, E. (2023), 'Last-Mile Drone Delivery: Past, Present, and Future', *Drones*, 7(2): 77.

Fertner, C. and GROßE, J., (2016), 'Compact and resource-efficient cities? Synergies and trade-offs in European cities', *European Spatial Research and Policy*, 23(1): 65–79.

Goetz, K.S. (2010), 'Encouraging sustainable business practices using incentives: a practitioner's view', *Management Research Review*, 33(11): 1042–1053.

Grant, A., Cordle M., Bridgwater E. E. (2020), 'Quality of Recycling Towards an Operational Definition' Available at: https://publications.jrc.ec.europa.eu/repository/bitstream/JRC122293/final_quality_of_recycling_report.pdf.

Green, E.D. (2014), 'The political economy of urbanization in modern Botswana', *SSRN*, Available at: https://papers.ssrn.com/sol3/papers.cfm?abstract_id=2251935.

Hussain, S. and Sheikh, H. (2023), 'Sustainability Assessment by Green Building Rating Systems: A Comparative Analysis of LEED, BREEAM and CASBEE on a Case Study', *International Journal for Multidisciplinary Research*, 5(6): 1–29.

James, N. (2024), 'Urbanization and its impact on environmental sustainability', *Journal of Applied Geography Studies*, 3(1): 54–66.

Jelti, F., Allouhi, A. and Tabet Aoul, K.A. (2023), 'Transition Paths towards a Sustainable Transportation System: A Literature Review', *Sustainability*, 15(21): 15457.

Kampragou, E., Lekkas, D. and Assimacopoulos, D. (2010), 'Water demand management: Implementation principles and indicative case studies', *Water and Environment Journal*, 25(4): 466–476.

Kothari, R., Tyagi, V. V. and Pathak, A. (2010), 'Waste-to-energy: A way from renewable energy sources to sustainable development', *Renewable and Sustainable Energy Reviews*, 14(9): 3164–3170.

Kubanza, N.S. (2020), 'The role of community participation in solid waste management in Sub-Saharan Africa: A study of Orlando East, Johannesburg, South Africa', *South African Geographical Journal*, 193(1): 1–14.

Lawrence, M. (2015), 'Reducing the environmental impact of construction by using renewable materials', *Journal of Renewable Materials*, 3(3): 163–174.

Lem, L. (2002), 'Promoting economic development by improving transportation infrastructure for goods movement', *Jack Faucett Associates*, Available at: https://citeseerx.ist.psu.edu/document?repid=rep1&type=pdf&doi=77f7171898894357f860a926297848f56c8f1b7f

Lepczyk, C. A., Aronson, M. F., Evans, K. L., Goddard, M. A., Lerman, S. B. and MacIvor, J. S. (2017), 'Biodiversity in the city: Fundamental questions for understanding the ecology of urban green spaces for biodiversity conservation', *BioScience*, 67(9): 799–807.

McDonald, R.I., Aronson, M.F., Beatley, T., Beller, E., Bazo, M., Grossinger, R., Jessup, K., Mansur, A.V., Puppim de Oliveira, J.A., Panlasigui, S. and Burg, J. (2023), 'Denser and greener cities: Green interventions to achieve both urban density and nature', *People and Nature*, 5(1): 84–102.

Mittal, S. and Biswas, A. (2019), 'Exploring transportation demand management as a strategy to decongest Indian cities and improve mobility', *International Review for Spatial Planning and Sustainable Development A: Planning Strategies and Design*, 7(4): 185–211.

Morseletto, P. (2020), 'Targets for a circular economy', *Resources, Conservation and Recycling*. 153: 104553.

Ogunkunbi, G. A. and Meszaros, F. (2023), 'Investigating the role of urban vehicle access regulations as a policy tool for promoting electric mobility in Budapest', *Urban Science*, 7(2): 39.

Oloruntobi, O., Mokhtar, K., Rozar, N.M., Gohari, A., Asif, S. and Chuah, L.F. (2023), 'Effective technologies and practices for reducing pollution in warehouses-a review', *Cleaner Engineering and Technology*, 13: 100622.

Pardo-Bosch, F., Blanco, A., Sesé, E., Ezcurra, F. and Pujadas, P. (2022), 'Sustainable strategy for the implementation of energy-efficient smart public lighting in urban areas: Case study in San Sebastian', *Sustainable Cities and Society*, 76: 103454.

Perry, G. and Cox, R.D. (2024), 'Opportunities for biodiversity conservation via urban ecosystem regeneration', *Diversity*, 16(3): 131.

Pietrucha-Urbanik, K. and Rak, J., (2023). Water, resources, and resilience: Insights from diverse environmental studies. *Water*, 15(22), p. 3965.

Price, J. L. and Joseph, J. B. (2000), 'Demand management – A basis for waste policy: A critical review of the applicability of the waste hierarchy in terms of achieving sustainable waste management', *Sustainable Development*, 8(2): 96–105.

Rajani, R.L., Heggde, G.S., Kumar, R. and Chauhan, P. (2022), 'Demand management strategies' role in sustainability of the service industry and impacts on company performance: using SEM approach', *Journal of Cleaner Production*, 369: 133311.

Rudnick, D., Ryan, S. J., Beier, P., Cushman, S. A., Dieffenbach, F., Epps, C., Gerber, L. R., Hartter, J. N., Jenness, J. S., Kintsch, J., Merenlender, A. M., Perkl, R. M., Perziosi, D. V. and Trombulack, S. C. (2012), 'The role of landscape connectivity in planning and implementing conservation and restoration priorities. Issues in Ecology', *University of New Hampshire Scholar's Repository*, Available at: https://scholars.unh.edu/cgi/viewcontent.cgi?article=1018&context=geog_facpub.

Saad, R., Plazas-Niño, F., Cannone, C., Yeganyan, R., Howells, M. and Luscombe, H. (2024), 'Long-term energy system modelling for a clean energy transition and improved energy security in Botswana's energy sector using the open-source energy modelling system', *Climate*, 12(6): 88.

Sánchez-García, E., Martínez-Falcó, J., Marco-Lajara, B. and Manresa-Marhuenda, E. (2024), 'Revolutionizing the circular economy through new technologies: A new era of sustainable progress', *Environmental Technology & Innovation*, 33: 103509.

Vega-Gonzalo, M., Gomez, J., Christidis, P. and Vassallo, J.M. (2024), 'The role of shared mobility in reducing perceived private car dependency', *Transportation Research Part D: Transport and Environment*, 126: 104023.

Waseem, M., Mutahir Ullah Ghazi, S., Ahmed, N., Ayaan, M. and Kebede Leta, M. (2023), 'Rainwater harvesting as a sustainable solution to cope with drinking water scarcity and urban flooding: A case study of public institutions in Lahore, Pakistan', *CivilEng*, 4(2): 638–656.

White, K., Habib, R. and Hardisty, D. J. (2019), 'How to SHIFT consumer behaviors to be more sustainable: A literature review and guiding framework', *Journal of Marketing*, 83(3): 22–49.

Xu, X., Watts, A. and Reed, M. (2019), 'Does access to the internet promote innovation? A look at the US broadband industry', *Growth and Change*, 50(4): 1423–1440.

Yazdani, S. and Lakzian, E. (2023), 'Conservation; Waste Reduction/Zero Waste', in D.S.K. Ting, J.A. Stagner (eds), *Pragmatic Engineering and Lifestyle*, 131–152, Leeds: Emerald Publishing.

Yigitcanlar, T. & Kamruzzaman, M. D. (2019), 'Smart Cities and Mobility: Does the Smartness of Australian Cities Lead to Sustainable Commuting Patterns?' *Journal of Urban Technology*, 26(2), 21–46, April. DOI: 10.1080/10630732.2018.1476794.

Chapter 9

UNLOCKING THE VALUE FROM COFFEE WASTE: A CASE STUDY OF SUSTAINABLE PRACTICES OF A COLOMBIAN COFFEE FARM

Claudia García Flórez

Introduction

The journey of coffee, a traditional delight cherished across continents, leaves behind a trail of waste that challenges our notions of responsible cultivation. It is estimated that 2.25 billion cups of coffee are consumed worldwide every day (International Coffee Organization, 2021). The global coffee industry generates approximately 23 million tonnes of waste annually, including coffee grounds, husks and other by-products. Colombia, a major contributor to this industry, produces around 14 million sacks yearly, equivalent to roughly 840,000 metric tonnes of coffee. Despite the prominence of Colombian coffee on the world market, many production practices remain traditional, on small coffee farms owned by local farmers. Many of these practices continue to use conventional and unsustainable methods such as chemical fertilisers and pesticides, while coffee by-products are often burned and disposed of. However, there is a growing interest in adopting regenerative and circular practices to produce higher-quality coffee. This shift helps to promote a more competitive and appealing product that meets the demands of customers who are increasingly looking for products with a fair and sustainable origin, ultimately leading to greater profitability for both customers and producers.

This chapter looks at coffee production and investigates the potential uses of the waste generated during the process. The field visit took place at the El Eden farm (Figure 10.1), located in the heart of Guaduas, La Carbonera district, a north-western municipality in the Cundinamarca region of Colombia. This area has a temperate climate ranging from 23.5°C to 28°C and is characterized by its mountainous landscape. Efforts are being made here to reuse coffee by-products, transforming them into valuable resources. This chapter explores the recycling processes occurring on the coffee plantation and looks at alternative methods for developing new materials using these by-products. It provides insights into how future research can drive innovation, reduce carbon footprint, and

Figure 9.1 Coffee farm El Eden.

promote sustainable practices through regenerative agriculture. Additionally, these advancements have the potential to generate positive socio-economic impacts for local farmers, who are just beginning to explore the underutilized value of waste and by-products beyond coffee production.

Within this coffee plantation, a commitment to sustainability goes beyond the surface. By-products of coffee production, including pulp and parchment, are repurposed as integral components in the production cycle. The pulp is reintegrated into the plantation's production process, enriching the land as organic fertilizer. Meanwhile, the parchment, with its calorific energy, plays a crucial role in coffee drying and the formulation of eco-friendly pesticides. This approach exemplifies the transformation of waste, where innovation meets tradition and sustainability intertwines with agriculture, potentially opening new lines of business for farmers.

Case Study in El Eden Farm

To conduct the case study, I chose the El Eden farm in Guaduas, Cundinamarca. This selection was driven by the region's notable agricultural expansion, particularly in coffee cultivation. Additionally, I considered the producers' keen interest in evolving cultivation techniques that go beyond merely utilizing coffee by-products. They are actively exploring methods to enhance soil quality and crop yield, harnessing the properties of what is traditionally perceived as waste. This initiative is the first step in the study of these by-products to develop innovative materials. During the field visit, the participating farmers were interviewed, offering valuable insights into the coffee production process. This encompassed the identification of waste generated at every stage, from the harvesting of coffee cherries to the preparation and cultivation of crops. The chapter has identified current methods of utilizing coffee by-products and waste. Within the production process, two primary types of biowaste were identified for reintegration: pulp and parchment. Following the identification of the waste generated not only on the farm but also in the post-consume stage, a proposal was made to initiate an experiment focusing on parchment and used coffee grounds to assess the development of a novel material. The first section examines the use of coffee parchment. This is a thin and silvery layer that envelops the coffee beans within the cherry. It serves as a protective barrier as it provides some defence against moisture and external elements. This by-product is of considerable value due to its calorific properties, which are derived from the bean once it has completed the drying process. As a result of these characteristics, coffee parchment is used as a fuel source for coffee dryers. Furthermore, the calorific capacity of this by-product is exploited by the elaboration of a sulphur-lime composition. This is achieved by the mixing of the coffee parchment with rice husk and burning organic material waste (e.g., bones, plants, and agricultural by-products). This process results in the production of rich mineral ash, which is subsequently mixed with water and used in the prevention of phytosanitary issues, as well as the mineral enrichment of the soil in coffee crops. Two distinct applications were explored. The initial approach involved utilizing the parchment as a biomass fuel. The second method consists of forming a clay-like substance by blending the parchment with waste coffee grounds, which was then used to test the preliminary elaboration of a novel material. To create the mixture, 80 per cent of the dry coffee ground waste and 5 per cent of the parchment (broken into small pieces) were mixed in a container with natural binders (15%) such as starch, guar gum and xanthan. This integration facilitates the binding and shaping of the materials. Water was gradually added to the process, ensuring the homogeneity of the mixture, which is essential for shaping. After the clay mixture was obtained, it was added to the moulds. Tools such as a press can be employed to compact and ensure the steadiness of the mixture within the moulds. The containers were set for an air-drying phase: a period of three days at room temperature (16-18°C) was required make sure the piece was completely dry. This period may vary depending on the dimensions of the containers and the prevailing moisture levels, which are calculated based on the initial and final weight of the mix. The application of pressure through a press during the drying process increased both density and strength, and it is an important step to ensure the correct agglomeration of the particles in the mortar. After drying and curing, the pieces were carefully removed from the moulds. A final agglomerant, consisting of glue made of corn and casein, was applied to the surface of the moulded mix with a trowel to create a plasticized surface and provide edge smoothing.

Results and Discussion

Coffee Parchment Use

The use of coffee parchment has resulted in an efficient ignition process with a calorific power of 17 MJ/kg to supply heat to the drying equipment; it is very efficient in producing the sulphur-lime composition by generating ash from organic material (Figure 10.2). It is beneficial not only for its calorific capacity but also for its physical-chemical properties that give a higher nutritional quality to the mixture. This prevents phytopathology issues, as well as enriching the soil (Cabrera, et al., 2018). Coffee parchment, although representing only 4.2 per cent of fresh fruit, is very efficient when used as a fuel, even more so than rice husks.

Coffee Grounds and Parchment Mixture

From the prepared and moulded coffee clay mixture, samples of an agglomerated material were obtained (Figure 10.3 (a)). The pieces consisted of a 4mm plate, light in weight and with a soft, porous texture. The porosity varies according to the coffee grind: the lower the grinding size, the lower the porosity and the higher the cohesion of the mixture. The coffee-derived clay displays an inherently viscous,

Figure 9.2 Coffee parchment use: (a) for the dryer; and (b) mixing organic ingredients for a sulphur-lime composition.

Figure 9.3 (a) material (coffee sheet) from the coffee clay mixture; (b) texture without agglomerate addition on the surface; and (c) texture with agglomerate agent applied on the surface.

malleable and adhesive texture attributed to the incorporation of compacting agents. The coffee grounds contain lignin, which serves as a 'glue' and a reinforced material compatible with starch-based binders, primarily sourced from corn, along with guam gum which has antioxidative properties (Zarrinbakhsh et al., 2016). This creates a kind of bio-based composite, which is improved mechanically with the support of coffee parchment.

The final product, after drying, is a light-brown substance with a consistency similar to plaster. The texture of the material (coffee sheet) could vary depending on the agglomerant agent applied on the surface (Figure 10.3(b)(c)). Although the material is at an early stage of development, it is anticipated that further improvements will be made to evaluate its potential for use in the elaboration of biodegradable containers for food use, plant containers and agglomerates to produce pieces as a substitute for wood, construction material or for handicraft use. The granulated coffee residue has already been studied for use in various products, such as cups, composites (Zarrinbakhsh et al., 2016), sound-absorbing panels (Yeol et al., 2020), ceramics for construction (Vilao & Simao, 2014), and textiles (Kumar, 2023), among others.

Conclusion

During the visit to the El Eden farm, During the visit to El Eden farm, innovative approaches for utilizing coffee production residues – specifically the husks and cherry pulp – were identified. These by-products were integrated into the same coffee cultivation following the principles of the circular economy. The farmers have observed that by utilising the caloric energy of the hulls and the nutritional contributions of the pulp, they have been able to achieve notable outcomes in terms of soil and plant behaviour. Thus, using coffee by-products supports the transition toward regenerative and more sustainable practices in coffee production. Three coffee by-products were studied for this research: parchment, cherry pulp, and coffee grounds remaining after brewing. The first two residues were already used on the farm to produce non-chemical fertilizers and fungicides. After the visit, experiments were carried out using parchment and residual coffee grounds, resulting in a first sample (coffee sheet) of a potential material made from a kind of coffee clay or bio-composite.

As a first step in testing a novel material, it is necessary to continue the development of new experimental versions of the sheets made from coffee clay. In-depth laboratory analyses of the materials are highly recommended to evaluate additional physical-mechanical properties, such as strength and flexibility. Based on this evaluation, suggestions for improvements can be made, new mixtures can be explored and applications can be identified for creating products from these sheets. It is crucial to ensure the use of sustainable-sourced materials and practices while also ensuring the biodegradability of the final product.

Based on the results obtained, it would be valuable to experiment with different binding agents, such as cellulose nanofibers, starch from various sources, or by-products from the same farms, like cassava or yam. Additionally, plasticizers such as glycerol and sorbitol have been documented to enhance flexibility and improve the filling properties of these materials when combined with certain bindings. Further research in this field to improve this first assay and develop new products from coffee waste may provide additional sources of income for the farmers and enhance the bio-economical potential in the region.

Acknowledgements

This acknowledgement is extended to El Edén farm owners the Puentes Ruíz family, whose cooperation and willingness have enabled a thorough exploration of sustainable coffee cultivation. Their contribution has enriched this chapter with an invaluable perspective.

References

Cabrera, M., Robledo, J. and Soto, A. (2018), 'Actividad insecticida del caldo sulfocálcico sobre Hypothenemus hampei (Coleoptera: Curculionidae). *Boletín Científico Centro de Museos. Museo de Historia Natural,* 22(2): 24–32.

Dzung, N.A., Dzung, T.T. and Khanh, V.T.P. (2013), 'Evaluation of coffee husk compost for improving soil fertility and sustainable coffee production in rural Central Highland of Vietnam', *Resources and Environment,* 3(4): 77–82.

International Coffee Organization (2021), 'Coffee Market Report'. Available at: http://www.ico.org/new_historical.asp?section=Statistics.

Iriondo, A., DeHond, M. and Castillo, M. (2020), 'Applications of Compounds from Coffee Processing By-products', *Biomolecules,* 10(9): 1219.

Kumar, A. (2023), 'Coffee fibres from coffee waste', in R. Nayak (ed), *Sustainable Fibres for Fashion and Textile Manufacturing,* 287–307, Cambridge: Woodhead Publishing.

Sena da Fonseca, B., Vilão, A., Galhano, C. and Simão, J. A. R. (2014), 'Reusing coffee waste in the manufacture of ceramics for construction', *Advances in Applied Ceramics,* 113(3): 159–166. doi: 10.1179/1743676113Y.0000000131.

Yeol, B., Mi Cho, H., Sung, C. and Berardi, U. (2020), 'Circular reutilization of coffee waste for sound absorbing panels: A perspective on material recycling', *Environmental Research,* 184: 109281.

Zarrinbakhsh, N., Wang, T., Rodriguez, A., Misra, M., & Mohanty, A. (2016). 'Characterization of wastes and coproducts from the coffee industry for composite material production'. *Bioresources*, 11, 7637–7653. https://doi.org/10.15376/biores.11.3.7637-7653

Part III

DESIGN AND BUSINESS

Chapter 10

Scaling Up Upcycling: A Comparative Analysis of Furniture, Lighting and Bags Made Through Repurposing

Patrick Pearce, Martin Racine, Isabel Ordóñez and Alice Jarry

Introduction

The upbeat feel of the word 'upcycling' makes it a handy catchphrase in academic research, the business world and DIY culture alike for embracing repair, refurbishing, remanufacturing, reuse and repurposing, as well as certain forms of recycling. This chapter focuses on upcycling through repurposing, also known as creative reuse, in the sectors of furniture, lighting and bag design, and how this practice can be scaled up. The furniture sector has been identified as a critical sector for circular economy initiatives, as it generates large amounts of waste yet typically operates with small-scale, flexible production systems (European Commission, 2020; Oakdene Hollins and Dillon, 2021). Thus far, the sector's upcycling-related activities have been predominantly carried out by individual artisans or micro-scale enterprises focusing on resale, repair and refurbishment. A handful of more organized small, medium and large enterprises underscore the sector's circularity potential, notably through the remanufacturing of office furniture. However, upcycling research in furniture remains scant.

Despite its advantages, upcycling remains a niche, small-scale practice in developed countries (Singh et al., 2019). Scaling up upcycling offers designers the possibility of working sustainably, meaningfully and viably while increasing material circularity. The problem of scaling up upcycling is vast and research in recent years has examined a wide range of interconnected challenges and enablers, from macro and meso-level government and systemic perspectives to micro-level, designer and individual enterprise aspects. A few case studies on upcycling enterprises have shed light on specific business and operational aspects relating to scaling up. This research project offers a comparative analysis to expand on these cases and draw broader conclusions to help improve the possibility of scaling up among designers practising upcycling through repurposing. Repurposing has been defined as a form of reuse of discarded material for a new purpose, which brings about 'new value creation with a minimum loss of material integrity' (Van Hees & Oskam, 2021, p. 2). This project further clarifies repurposing as the use of waste objects or material, whose original form and structure are still recognisable, for a different function than what they were initially produced for while maintaining this recognisability and increasing value in the resulting new object or material.

In terms of an enterprise's day-to-day practices, scaling up involves two key dimensions: first, the scalability of the organization and its operations – an organization's ability to grow with increased production without being constrained by its structure or available resources and second, the viability of the business or a project – its ability to carry on and sustain profits over the long term. To understand scalability and viability as they apply to upcycling enterprises, a cross-disciplinary analysis allows us to dimensionalize the connection between upcycling and craft and to examine the ways in which upcycling creates value and how consumers perceive this value.

Upcycling, Craft and Scalability

The scalability of upcycling is constrained by its craft nature, a characteristic identified in previous research on upcycling (Singh et al., 2019; Sung et al., 2021; Sung & Cooper, 2015; as cited in Sung et al., 2022). While upcycling has resulted in successful small and medium-sized (SME) enterprises, the majority of professional upcyclers can be described as designer-makers, a term increasingly used to refer to craftspeople (Luckman & Andrew, 2020). Upcycling and craft share positive ethical qualities (Archer & Björkvall, 2021; Zhan & Walker, 2019); given its environmental and social benefits, upcycling can be seen as a form of sustainable craft. In terms of production benefits, craft has been found to offer the flexibility to scale up or down according to project needs. Craft's interdisciplinary nature also helps to solve design and production problems, and craftsmanship adds value to design and its communications (Woolley, 2011).

While craft's advantages have the potential to help benefit repurposing projects, upcycling and craft also share issues of production efficiency and viability (Caldera et al., 2022; Zhan & Walker, 2019) which, along with the complexity of working with waste objects and materials, may well explain the time-consuming and labour-intensive processes of upcycling (Singh et al., 2019; Sung & Abuzeinab, 2022). Other key scalability challenges facing upcyclers on a day-to-day operational level include access to materials, affordable space, equipment and expertise (Singh et al., 2019). Furthermore, each industry has particular technical and organizational challenges and a need for tailored solutions (Caldera et al., 2022). Furniture production materials and finished products can require significant storage space, and certain materials require specialized expertise to work with. In a nutshell, the craft nature of upcycling through repurposing is a major stumbling block to production scalability. While not all designer-makers or upcyclers desire growth (Luckman, 2018; Sung et al., 2022), upcycling requires some level of scale to help its practitioners earn a viable living. It is striking to note that during the search for upcycled products to be included in this analysis, a significant number of promising examples were found to be no longer available.

Upcycling Value Creation, Purchase Drivers and Viability

The viability of a business or project is related to multiple factors, including the managerial abilities of the individuals involved, access to capital and other resources, strategic partners and cost structure. In light of research findings and resources in the field of circular business model canvases, this research focuses on elements of viability that can be evaluated with relative ease based on secondary source information, namely product value proposition, distribution channels, communications and community building and environmental and social sustainability factors (Lewandowski, 2016; Live Circular Canvas Consortium, 2020).

The value created by upcycling significantly influences the perceptions and desirability of upcycled products. Upcycling enhances value through design, creativity, and storytelling, transforming discarded items into objects with new functional or aesthetic appeal. This process generates four key forms of value:

- *Design/Creativity Value*: through the creative idea of its new purpose or context, through unique aesthetics, and or through enhanced functionality or innovative features making them attractive to consumers (Baik & Kim, 2014; Buck & Lee, 2020; Yu & Lee, 2019).
- *Environmental Sustainability Value*: through the diversion of waste from landfills and offering an alternative to products made from virgin materials (Archer & Björkvall, 2017).
- *Ethical Value*: through messages of sustainability and anti-consumerism, reflecting ethical consumption practices and promoting social responsibility (Adıgüzel & Donato, 2021; Archer & Björkvall, 2017).
- *Storytelling Value*: through the story of the object's transformation, which enhances the consumer's connection and engagement with the object. This occurs inherently when the object's original identity is recognisable, a concept known as 'past identity salience' (Kamleitner et al., 2019), which can also be conveyed through mediums such as websites, packaging or other forms of communication to increase audience engagement and sales (Kamleitner et al., 2019).

These forms of value feed into the value proposition of the upcycled object in various proportions, affecting its perceived ability to meet people's needs and thus influencing its sales potential and viability. With repurposing, the transformation story is inherently greater than with repair, refurbishing or remanufacturing, so we tend to see repurposed objects as more creative or unique, and their story is more compelling. On the other hand, upcycled products can also be perceived by consumers as risky, with functional, social and financial risks identified as critical barriers to purchase (Park & Lin, 2020; Yoo et al., 2021; Yu & Lee, 2019). Being upcycled is no guarantee of desirability and some upcyclers prefer to focus on aspects other than the upcycled nature of their designs (Singh & Ordoñez, 2016). In a broader market perspective, the ability to compete with non-upcycled products has been identified as a critical success factor in upcycling enterprise case studies (Sung et al., 2022).

As part of a broader research project on improving scalability and viability in upcycling design, this research aimed to expand on previous case studies and help fill the knowledge gaps in scaling up upcycling enterprises (creating furniture, lighting and bags) from an operational scalability and business viability perspective. While the focus was on designer-driven micro and small enterprises, one medium-sized and two large enterprises were included in the analysis. The results, therefore, may be of interest to larger design studios or established manufacturers considering upcycling projects.

Methods

To investigate the issue of scaling up upcycling beyond individual case studies, this research employed comparative analysis. Commonly employed methodology in the social sciences when data is insufficient to enable statistical analysis, comparative analysis seeks causal relationships across two or more cases, with some level of explanation beyond simple description (Pickvance, 2001).

Sample and Comparative Analysis Design

The cases of upcycling through repurposing waste materials to create new furniture and bags (n=20) were discovered through Internet research. To ensure a variety among the 20 cases included in the study, four sub-categories were selected: seating, small tables, lighting and bags. Although not related to furniture or interior design, bags were included to open perspectives on scaling up. From the initial search results, projects were selected based on their apparent product quality and craftsmanship and a minimum of five years of business operation as an indication of the relative viability of the enterprises behind them. Additionally, the selection aimed for a variety of waste materials, price points, and production strategies. The following tables (Tables 11.1–11.3) present material scalability factors, production scalability factors, and product and business viability factors considered in the comparative analysis, developed on the basis of literature review findings and insights.

Data Collection

Internet research was carried out using Google as well as Pinterest, Instagram and Facebook and key words 'upcycled', 'upcycling', 'repurposing', 'reuse' and 'waste' in combination with 'design' and key words related to the furniture, lighting and bag categories. Case data gathered was predominantly secondary source, with information

Table 10.1 Material scalability factors

Scalability Factors	Analysis Criteria	Categorical Values
Waste material	Description of waste material used in the object	–
Waste material uniformity	Degree of uniformity in waste material form	high, medium, low
Waste source(s)	Description of material source(s)	–
Waste incidence	Qualification of the rate of occurrence of the waste material across its various sources	sporadic, periodic, regular, continuous

Table 10.2 Production scalability factors

Scalability Factors	Analysis Criteria	Categorical Values
Production strategy	General strategy used in the production system	Periodic batches, batch-to-order, made-to-order, custom batch-to-order, custom made-to-order
Design and production approach	Design and production characteristics and key steps in material preparation and production	–
Product options	Size, colour or other customization options	–
Production complexity	Steps involved, labour intensity, level and prevalence of required skills	high, medium, low

Table 10.3 Product and business viability factors

Viability Factors	Analysis Criteria
Design/creativity value	Uniqueness, innovation, functionality, quality, features, aesthetics, product options
Storytelling value	Past identity salience of the object, design or communication initiatives to convey the story of its past identity
Environmental sustainability value	Material that is difficult to recycle, the proportion of upcycled vs new material content of the object, other sustainability practices of the enterprise
Social sustainability value	Socially responsible production, profit sharing
Price competitiveness	As compared to products of comparable functionality, quality and design/creativity value on the market (upcycled or not)
Distribution channels	Self-distribution, number and complementary nature of other distribution channels, shipping factors
Communications and community building	Community of interest related to the previous life of the material or resulting product, social media approach, quality and consistency
Number of followers	Number of followers on Instagram
Staff size	Number of full-time employees, including the owner, on average
Business started	Year the company or individual started doing their business

gathered on websites and social media channels associated with the enterprises supplemented at times with existing articles and case studies. Emails were also exchanged with designer-makers or other staff in the enterprises to validate the information gathered, particularly regarding material sourcing and production steps.

Data Analysis

Based on the production approach, labour intensity, and the skill level required, an assessment of production complexity was evaluated for each case, with three possible values of high, medium or low. This categorical assessment was qualitative and was not validated by the enterprises. For instance, when the process required relatively rare skills related to the designer's tacit knowledge, (such as metal forming techniques using bicycle rims) and/or high labour intensity and multiple handmade processes, complexity was rated medium or high. Other qualitative factors have been synthesised as:

- Scalability enablers and scalability challenges, relating to both material and production aspects of scaling up from an operational standpoint; and
- Viability enablers and viability challenges relating to the project and the business, such as its distribution network, brand reputation, and so on.

The secondary data gathered and subsequent validations with the upcycling designers or companies were summarised for seating, tables, lighting and bags in the comparative analysis tables, which can be accessed through the online appendix at: https://spectrum.library.concordia.ca/993978/1/Upcycling%20comparative%20analysis_dataset.pdf

Results

Furniture and Object Case Characteristics

The cases selected resulted in a wide variety of products within each of the four categories and particularly across the 20 projects, ranging from pendant lamps serially produced from wine bottles, priced at €49, to a complex coffee table hand-crafted to order from scrap wood and designed by upcycling design pioneer by Piet Ein Eek, with a price tag of €3,014. Thirteen cases were micro-enterprises of one to nine employees, including five solo entrepreneur ventures (who, in some cases, outsourced to production partners). Five other cases were well-established small enterprises of 10 to 49 employees. Freitag, with 250 employees, is the largest upcycling-driven manufacturer. In the large enterprises cases of Delta Light and Lufthansa subsidiary Miles & More, the upcycled product was part of a special project within the companies' typical product lines. The data for the 20 projects covered 21 enterprises, as Miles & More, the producer, outsourced manufacturing to upcycling studio B2L Group, so information from both companies was included.

Table 11.4 provides a summary of categorical data aggregated from the projects/cases detailed in the online appendix. In half of the cases, the waste material was from commercial or institutional (public service-related) sources. Four cases used industrial (manufacturing) waste, while the remaining seven cases used post-consumer waste. Waste material uniformity was assessed as either medium or high, with waste incidence rates that were regular or constant for 16 of the 20 cases.

Periodic batch production was predominant, accounting for two-thirds of the cases. Danish enterprise Traels, which only sells to institutions or businesses, uses

Table 10.4 Upcycling projects categorical data summary

Factors	Categorical values	Seating	Small tables	Lighting	Bags	Total cases
Waste source type	Post-consumer	4	–	1	2	7
	Industrial	–	1	2	1	4
	Commercial/Institutional	1	4	2	3	10
Waste uniformity	High	2	3	2	3	10
	Medium	3	2	3	2	10
	Low	–	–	–	–	–
Waste incidence	Sporadic	–	1	–	1	2
	Periodic	1	1	–	–	2
	Regular	4	3	3	4	14
	Constant	–	–	2	–	2
Production strategy *	Periodic batches	1	3	5	5	14
	Batch to order only	1	–	–	–	1
	Batch + made to order components	2	–	–	–	2
	Made-to-order only	1	2	–	–	3
	Custom batch to order option	1	–	–	3	4
	Custom made-to-order option	2	1	1	–	4
Production complexity	High	2	3	1	2	7
	Medium	3	2	1	1	8
	Low	–	–	3	2	5
Number of followers	< 1,000	1	3	1	–	5
	1,000 to 3,000	3	1	–	–	4
	3,000 to 10,000	1	–	2	3	6
	10,000 +	–	2	1	2	5
Staff size*	1	2	1	2	–	5
	2 to 9	2	2	2	2	8
	10-49	1	2	–	2	5
	50+	–	1	1	1	3
Years in business *	5 to 10	3	3	4	–	10
	10 +	2	3	1	5	11

* One of the projects involved two companies, so the company-related data sample totals 21 instead of 20

a 'batch-to-order' strategy, requiring a minimum order of ten units of its Airbag puff cushion. Two seating projects adopted a mixed approach, with the structure of the stool or chair produced in batches and their seats (and backrest in one case) made to order, allowing for custom colour or artwork options. Eight of the 20 cases also offered custom order options for individuals or multi-unit corporate orders.

As a critical scalability factor, the results of the production complexity assessment underscore the craft production nature of repurposing, with only five of the 20 projects evaluated as low complexity and the rest evenly divided between medium and high complexity. As a potential indicator of viability, fewer than half the enterprises had less than 3,000 Instagram followers at the time of writing, and five had less than 1,000, which may also suggest that these enterprises may be connecting with their audiences in ways other than Instagram. Six cases had between 3,000 and 10,000 followers, while the five remaining were over 10,000, with bag-makers Elvis & Kresse at over 35,000 and Freitag at over 192,000, numbers which also reflect the longevity of these companies.

Material Selection and Sourcing Scalability Factors

Decisions related to waste material and its sources are key to both scalability as well as viability, as they have a direct impact on costs. Table 11.5 summarizes enablers and challenges in this regard. Material prevalence is an obvious factor. Abundant materials such as wine bottles and scrap wood are commonly seen in

Table 10.5 Material selection and sourcing scalability enablers and challenges

Material scalability factors	Enablers	Challenges
Waste material selection	Prevalent or abundant material	Limited, uncommon or rare material
	Foreseeable prevalence in the future	Uncertain availability in the future
	Material with limited lifespans	Material that is becoming outdated
	Difficult to recycle material	Material with a high value for recyclers
Uniformity	High uniformity	Low or medium uniformity
Source(s)	Multiple sources	Single source or only a few sources
	Close/easy-to-access sources	Far/difficult to access sources
	Reliable source	Unreliable or difficult to plan around sources
	Establish one's own sourcing network	
Incidence	Occurs regularly or constantly	Single opportunity
		Occurs sporadically or periodically
Transportation	Low transportation costs	Significant or high transportation costs

upcycled products; however, this can result in increased competition and less product differentiation, thus affecting viability. Cardboard pendant lamps are a good example of this. The relative difficulty of sourcing less common materials such as skateboards or used sails serves to illustrate the advantages of establishing sourcing networks. For example, Alchemy Goods installs branded bins in a network of bike shops to collect inner tubes, as does Mister Wilson's with skate shops to collect used and broken skateboards.

The cases here show that commercial sources can work as regular sources for materials such as discarded office chairs, truck tarpaulins and wood offcuts. Institutional sources such as fire brigades (fire hoses) and national aviation or train companies (multiple components) are also reliable, with decommissioned materials available on a periodic basis. In contrast, manufacturers can offer a range of materials, from offcuts to non-compliant products. Material with inherently limited lifespans can offer good opportunities, too. Lifespan can be limited by periodicity, such as auction banners, by decreasing performance over time, such as sails or truck tarpaulins and by safety standards, such as fire hoses.

However, material transportation and storage costs and environmental impact should also be considered. For example, large sails weighing 400 to 600 kg require a platform truck to move and sources may be located far from the production location. Material selection also impacts production: time and costs required to prepare particular materials for reuse, such as cleaning, repair, disassembly, and cutting to sizes manageable for production. For example, to extract specific components during the dismantling of Lufthansa aircraft, Miles & More collaborates with a partner company that possesses the requisite machinery and technical expertise.

Material Transformation, Design and Production Scalability Factors

The processes, resources, partners and strategies involved in transforming waste material into furniture and object designs all influence production scalability, which in turn affects business viability. Table 11.6 offers an overview of how these factors work. The material form can become a production scalability opportunity: if the form is already engaging and functional for the new purpose, it may only require minimal transformation. Delta Light's aluminium profile offcuts are a good example, with their sophisticated extruded form requiring only simple cutting, grinding, powder coating and assembly to work as a very contemporary-looking wall sconce. Commonly available tools, facilities and processes requiring common skills, such as sewing and standard metalwork, help keep costs in line. For Rebottled's lamps made from wine bottles, the relative simplicity of the production approach also enables the enterprises to outsource their production to bring down their costs and adapt to demand fluctuations.

Automation is a significant enabler of production efficiency, although it typically requires the use of uniform materials. The flooring and ceiling panels from Dutch Railways decommissioned regional trains are uniform enough to use CNC routers to cut forms for Veldtwerk's chair and OpNieuw!'s office side table.

Table 10.6 Material transformation, design and production scalability enablers and challenges

Production scalability factors	Enablers	Challenges
Preparation for reuse	Material requiring little preparation	Material requiring significant preparation
Transformation	Material and design requiring minimal transformation	Material and design requiring significant transformation
Tools and facilities	Material and design requiring commonly available tools and facilities	Material and design requiring specialised tools and facilities
Processes and skills	Material and design that reduces production steps and time, and requires simple processes and/or common skills	Material and design that increases production steps and time, and requires complex processes and/or high level or specialised skills
Automation	Material and design that allows automation	Material and design that requires handmade processes
Material variations	Design that adapts easily to material variations	Design requiring significant changes for material variations
Manufacturing partners	Local, cost-effective manufacturing partners for major manufacturing steps or specialised processes	Long distances from manufacturing partners, or high local costs
Outsourcing	Material and design that allows cost-effective national or international outsourcing of production steps to manufacturing partners with equitable labour practices	Large, heavy materials or resulting products that impede outsourcing
Product options	No options or few options	Multiple options
Production strategy	Flexible batch production, batch-to-order, and/or limited series of significant size	One-offs, made-to-order, and/or small limited series

However, some upcycled products also incorporate a proportion of virgin materials to enable automation. Mister Wilson's stools are a good example, whereby the seat is made from used or broken skateboards and the legs are made from birch plywood, enabling them to be cut using a CNC router. Manufacturing templates such as cutting templates and sewing patterns used by bag makers or jigs used to hold steel parts together for welding help increase production efficiency and precision. Custom-made parts can be another enabler. Anti has 3D parts printed from recycled plastic, which, though relatively costly to produce, include brackets designed to work with all kinds of umbrellas and help keep overall production time to two to three hours per lamp.

Viability Factors

While scalability factors often impact viability as well, viability factors per se focus on the four types of upcycling value creation identified in the introduction, as well as price competitiveness, distribution, communications and community building. Table 11.7 aggregates viability findings from the cases. As with production scalability, material factors play an essential role in a product's viability. Some

Table 10.7 Viability enablers and challenges

Viability Factors	Enablers	Challenges
Design/creativity value	Design with unique, creative, innovative and/or aesthetic qualities	Many products use the same waste material
	Material variations leveraged to enable unique designs using existing templates	Unoriginal or outdated design/ aesthetics
		Overly niche design or aesthetics
	Material from rare sources or rare versions of materials to create limited editions	Trendy design becoming quickly outdated
		No colour options
Functional value	Material and design suitable for a variety of applications or environments	
	Product for multiple purposes or configurations	
	Features that increase functionality	
	A choice of colours, sizes or other options	
Storytelling value	Product design makes past identity easy to discern	Past identity is difficult to discern
	Product design includes text about past identity	Past identity is not communicated
	Packaging or merchandising mentions past identity	
	Website or social media refers to past identity	
	Connections to material's past life through details about its existence	

(*Continued*)

Table 10.7 (Continued).

Viability Factors	Enablers	Challenges
Environmental sustainability value	Material that is difficult to recycle A high proportion of material content is upcycled Non-upcycled materials are recycled Designed for long life Repair service to extend product life Platform for exchange or resale End-of-life take-back possibility	Significant proportion of new (virgin) material content Lack of information on product quality or longevity No repair option
Social sustainability value	Produced using equitable labour practices Produced through a social enterprise A proportion of profits donated to a social cause	Produced in a country known for unfair labour practices with no mention of these concerns
Price competitiveness	Competitively priced vs comparable value products (upcycled or not)	High prices vs comparable value products (upcycled or not)
Distribution	Distributed through different channels Many distributors are complementary to one another Only one unit of a product with a given material colour/ pattern made available at a time Low shipping costs	Self-distribution only High shipping costs
Community	Material or products with a community of interest	Inconsistent use of social media
Building and communications	Consistent social media posting to increase reach Engaging images of sourcing and production process Celebrating the past lives of material Celebrating the craftsmanship of the people involved High-quality photos of products in use contexts Participation in design and sustainability exhibitions Media coverage	Uninspired or low-quality content Low number of followers

materials come with an inherent community of interest, such as cyclists (Bike Furniture Design), sailors and sea-lovers (Salty Bag, and Dvelas' chair) or aviation enthusiasts (Lufthansa table). Material variations can be leveraged to help reinforce the uniqueness of a product: Freitag, for example, uses different colours and graphics of truck tarps to create unique designs. Materials from rare sources can create a sense of exclusiveness with a limited series, as does Mariclaro with leather from cult brands of vintage cars.

While uniqueness is often a strong driver of upcycled products, functionality and the ability to fit into a variety of applications can enable a broader market for a product. Simple white Dacron sail material makes Dvelas' outdoor chairs suitable for a wide range of environments, thus facilitating commercial sales. The stool that can work as a mini table, the laptop sleeve in a backpack, and the choice of sizes, colours, patterns, and finishes — all features which enhance the functionality, suitability and perceived value of an upcycled product. There are also different ways to leverage past identity in the storytelling of upcycled products. Markings occurring on the original material can be leveraged in the design, such as markings on a firehose that we see on the bag or typography on an aircraft fuselage that forms a pattern on a tabletop. Text relating to origins can be incorporated into the design of the product or its merchandising, as Mariclaro and Salty Bag do in different ways (see Figure 11.1). Each Salty Bag product includes numbers on its hang tags, which allow customers to find information on its website about the particular sail used in that bag: the type of sail, its geographic origins, and where it was sailed. Reinforcing uniqueness in their online store product selection, bag makers Mariclaro and Freitag websites appear to offer only one unit at a time of each bag model made with a given material source (colour and/or graphics). By adding a sales revenue stream, bag makers such as Alchemy Goods can produce multiple units branded with the client's logo (see Figure 11.2).

Figure 10.1 Past identity conveyed through product design and merchandising.

Left: Mariclaro Vie+ Purse Wow Airlines Airbus A321 (https://www.mariclaro.ca/collections/air-canada/products/mariclaro-vie-purse-wow-airbus-a321); middle: Salty Bag Instagram post 1 (https://www.instagram.com/p/CUhfYu2o30h); right: Salty Bag Instagram post 2 (https://www.instagram.com/p/BcC-XNCFvIM)

Figure 10.2 Corporate orders of alchemy goods products.

Left: Instagram post – Ever Thought About Making Your Promo Products from Old Bike Tubes?! (https://www.instagram.com/p/CLcQo4oAIj6/); right: Shimano Transforms Old Bike Tubes into Zipper Pouches (https://www.ecologicdesigns.com/shimano-transforms-old-bike-tubes-into-zipper-pouches/)

In terms of environmental sustainability value, some cases stand out through their use of materials that are difficult to recycle, such as umbrellas or fire hoses. Considering other sustainable practices, micro and small enterprises do not typically have the resources to organize practices such as end-of-life product take-back systems, but some offer extended repair guarantees. Elvis & Kresse and Freitag, two relatively larger enterprises in the sample, have been awarded accolades for a range of sustainability practices across their operations. Freitag notably offers free spare parts for self-repair and the possibility of swapping one's bag for that of another customer via the company's resale platform on its website. On the social sustainability front, Danish company Traels and Dutch enterprises Tolhuijs Design, OpNieuw! and Rebottled all employ workers from vulnerable sectors of society, either directly or through outsourcing partners. Moogoo's furniture production creates equitable employment in African countries such as Burkina Faso. Elvis & Kresse, a certified B-corporation and accredited Living Wage Employer, donates 50 per cent of their profits to inspiring charity causes.

Pricing is a strategic decision affecting sales and profit margins. In the bag market, products must compete with hundreds of others of similar form and functionality, and the pricing of the cases analysed tends to be competitive with prices of products that are comparable in quality. In the other categories, there were some variations, but the general sense of competitiveness with comparable products remained. Distribution was a weak point for many of the cases in the sample, with self-distribution through their website being the norm, fuelled by word-of-mouth and social media. Instagram can be a significant enabler of brand awareness, but this typically requires a large base of followers as well as a healthy level of follower engagement. A few cases have opted for more diversified distribution strategies, which can be an effective strategy if profit margins and resulting sales volume make it worthwhile. Three of the five bag distributors use third-party distributors, notably Alchemy Goods (165 retailers + several online stores + Amazon) and Freitag (30 company-owned stores + 257 independent retailers). In furniture and lighting, Dvelas and Tolhuijs Design also have networks of outside retailers or online stores. As large enterprises, Delta Light and Miles & More can leverage their international distribution networks to help find customers for their upcycling projects.

To build and nourish their communities of interest, the enterprises analysed employ a few key strategies. The first is to celebrate the past identity of materials with engaging photos that show these past lives in context, as does Bike Furniture Design and Dvelas, which notably uses stock photos and reposts of magnificent sails and sailboats from other accounts (see Figure 11.3). Another approach is to engage followers in the material sourcing and product production processes with images that also highlight the labour and craftsmanship involved and help justify the pricing. Freitag details its 15-step process on its website. Detailing the material composition of a product also underscores its sustainability in a transparent, credible way.

Figure 10.3 Instagram posts celebrating the past lives of material.

Left: Photo of a sailing yacht regatta (example of types of photo posted by Dvelas) (https://pixabay.com/photos/boat-veil-sailboat-sea-ship-5159223); right: Gregg, A. 1986. Photograph. Bike Furniture Instagram post – Campus usually has decent snowbanks. Greg Potvin on his Kos Cruiser at Lee Hall, Spring 1986 (https://www.instagram.com/p/BNztIQRhWy0)

Different mediums can serve different purposes. Videos on websites and social media provide an opportunity to present the underlying ethos and reveal the process that goes into making the products. Blog posts can tell a more detailed story of the design, production and features of specific products. In addition to website and social media content, Alchemy Goods employs online advertising to maintain the visibility of its products and to enhance the likelihood of conversion among visitors to its website. Given the competitiveness of the US bag market, this may well be an essential tactic for the company. Several enterprises have taken part in design or sustainability-related exhibitions, and many have benefited from media coverage by publications of various types, potentially helping to reinforce their authority and prominence in Google search results.

Discussion and Conclusion

Material and Production Scalability

In regards to material, the sourcing networks set up by many of the enterprises in this comparative analysis align with the case study findings of Sung et al. in terms of the effective use of material sourcing partnerships, which ensures the reliability and efficiency of waste material to work with (Sung et al., 2022). In terms of production, results from the 20 projects/cases confirm the prevalence of craft production techniques as a defining characteristic of upcycling based on repurposing. The companies Elvis & Kresse, Dvelas and Alchemy Goods have grown into significant small enterprises, while Freitag has become a serious medium enterprise with 250 employees and distribution network in 23 countries. Yet, the production of these companies continues to be characterized by craft methods. Two of them have had to resort to international production to meet their needs. Elvis & Kresse set up its second workshop in Turkey, while Freitag has outsourced sewing to lower-cost European countries. While internationalizing production has an impact on the environmental sustainability of their products, it seems to be a necessary trade-off to ensure a high enough level of scale for these enterprises. For smaller players, keeping production processes simple is an important theme in that it makes it easier to find workers with the skills required and keep costs relatively low.

Anti's use of 3D-printed parts to accelerate its umbrella lamp production is a good example of using technology to increase efficiency, dovetailing with previous case study findings of production innovation for competitive advantage (Sung et al, 2022). Automation using technologies such as CNC routing is another example, which works for uniform waste material. Design and creativity are integral to the value proposition of upcycling through repurposing. This leads upcyclers to behave in a manner similar to many other designer-makers of unique craft objects. It has been observed that these creators typically employ low-volume local and artisan production methods to create high-margin products that appeal to niche customer segments. These products are then sold via narrow distribution channels, such as direct online sales, local retailers, or specialized e-commerce platforms (Doussard et al., 2018). The findings confirm the prevalence of design and creativity in the cases and demonstrate how this business model can be implemented effectively. Uniqueness is achieved by leveraging the unique characteristics of the material, while the use of recurring designs, patterns, and jigs ensures efficiency.

Product/Business Viability

Although this research does not provide data on the financial viability of these projects/cases, the relative longevity of the companies suggests that most are at least surviving well enough thus far. Sung et al's (2022) paper presented three upcycling case studies that identified the success factors of competitiveness – one of which was competing with non-upcycled products in terms of quality and price. Based on comparable materials and handcrafted quality, the products appear to be competitively priced, especially in the bag segment. Previous case studies on Freitag and Elvis & Kresse highlight the success of these enterprises, which suggests that labour-intensive upcycled goods production in a competitive market (e.g., bags) can be viable with high enough prices. However, this can require an ability to position the products in the luxury category. As research on sustainability suggests, the broader environmental and social sustainability practices of these two companies help them stand out as exemplary and may well help make customers feel better about paying high prices. Freitag, as an early pioneer of upcycled bags, has become an icon over time, and for people visiting Switzerland, an iconic Swiss brand with an alternative vibe (Budgen, 2017).

Distribution models tend to reflect the relative scale of the enterprises. While most rely on self-distribution, a subset has demonstrated considerable success, as evidenced by their Instagram follower numbers. There appears to be a relationship between Instagram's success and the consistent, engaging quality of the posts. As a vehicle for the communication of design, creativity and storytelling about past identity and sustainability ethos, the effectiveness of social media in general, and Instagram in particular, appears to be a key viability enabler. This aligns with previous findings on marketing effectiveness (Sung et al., 2022; Sung & Abuzeinab, 2022). On the other hand, Traels and OpNieuw! show the potential for selling to business and institutional markets, at least as an additional channel to pursue, as other companies such as Rebottled and Dvelas seem to be doing.

Identity, Connections and Location Set the Agenda

For micro enterprises, scaling up is fundamentally shaped by founders' self-identity, their vision of their project and the resources at their disposal. Skater Florian Bürkle repurposes skateboards into stools, while cyclist Andy Gregg transforms bike parts into seating; each benefits from their communities for materal sourcing. Local partners play a crucial role in the success of sustainable business projects (Lewandowski, 2016; Sung et al., 2022), with waste material source partners being the first to align. Mark Howell's rainy London location makes lost or broken umbrellas an abundant waste material for him and the city's museums have been happy to provide him with a steady supply. After sourcing, building relationships with influencers or broader communities of interest provides paths for growth. For example, Dvelas founder Enrique Kahle is both a sailing enthusiast and an architect, giving him connections to each of those worlds and influencing the types of customers he has found as well as his production methods. Similarly, Markus and Daniel Freitag's background as cyclists informs Freitag's bag designs, and the company has always cultivated strong ties with this community through online engagement and live events, helping to give their brand its cult status. These elements collectively determine what upcyclers design, how they source materials, build networks, produce products, position themselves in the market, and scale their business.

Contributions, Limitations and Future Research

This comparative analysis of furniture, lighting and bags made by upcycling expands on previous case studies, enabling the identification of a wide range of scalability and viability challenges and enablers across 20 cases/projects (involving 21 companies). This research provides insight into several aspects of furniture and object upcycling from the perspective of both design entrepreneurs and enterprises. Cases were selected based on upcycling through repurposing, so results should be considered specific to this area of upcycling. Information gathered was mainly secondary data based on the websites, social media channels, and, in some cases, articles on the companies. Interpretation of enablers and challenges was carried out based on the primary author's experience and tacit knowledge. In light of the crucial role of scaling up in enhancing the impact of upcycling, primary source research among upcycling enterprises on scalability and viability could assist in the quantification of success factors and the identification of additional feasible paths to scaling up. Expanding the scope to encompass other sectors where repurposing is happening, such as fashion and architecture, and other forms of upcycling, such as refurbishing and remanufacturing, would provide invaluable insights into the scaling up of these sectors and upcycling approaches.

References

Adıgüzel, F. and Donato, C., (2021), 'Proud to be sustainable: Upcycled versus recycled luxury products', *Journal of Business Research,* 130: 137–146.

Archer, A. and Björkvall, A. (2017), 'The "Semiotics of Value" in Upcycling', in S. Zhao, A. Björkvall, M. Boeriis and E. Djonov (eds), *Advancing Multimodal and Critical Discourse Studies: Interdisciplinary Research Inspired by Theo Van Leeuwen's Social Semiotics,* 165–180, New York: Routledge.

Archer, A. and Björkvall, A. (2021), 'Discourses in and Around Upcycled Artefacts: A Social Semiotic Perspective' in: K. Sng, J. Singh, B. Bridgens, (eds), *State-of-the-Art Upcycling Research and Practice,* 29–32, Cham: Springer.

Baik, E. and Kim, J.H. (2014), 'A Study on the Trend Analysis of Upcycling Furniture Design', *Journal of the Korea Furniture Society,* 25: 112–121.
Buck, L. and Lee, S. (2020), 'Sustainable Design Approaches Using Waste Furniture Materials for Design Students', *Proceedings of the 22nd International Conference on Engineering and Product Design Education.* The Design Society.
Budgen, M. (2017), 'The Freitag brothers. From truck tarps to biodegradable fabrics, thinking in cycles since 1993', *Lifegate,* Available at: https://www.lifegate.com/freitag-brothers-interview.
Caldera, S., Jayasinghe, R., Desha, C., Dawes, L. and Ferguson, S. (2022), 'Evaluating barriers, enablers and opportunities for closing the loop through "waste upcycling": A systematic literature review', *Journal of Sustainable Development of Energy, Water and Environment Systems,* 10(1): 1–20.
Doussard, M., Schrock, G., Wolf-Powers, L., Eisenburger, M. and Marotta, S. (2018), 'Manufacturing without the firm: Challenges for the maker movement in three U.S. cities', *Environment and Planning A: Economy and Space,* 50(3): 651–670.
European Commission (2020), 'A new Circular Economy Action Plan For a cleaner and more competitive Europe', *European Commission,* Brussels.
Kamleitner, B., Thürridl, C. and Martin, B.A.S. (2019), 'A Cinderella story: How past identity salience boosts demand for repurposed products', *Journal of Marketing,* 83(6): 76–92.
Lewandowski, M. (2016), 'Designing the business models for circular economy—Towards the conceptual framework', *Sustainability,* 8(1): 43.
Live Circular Canvas Consortium (2020), 'Live Circular Canvas', Available at: https://livecircularcanvas.eu/livecanvas/LIve-Circular-Canvas.html
Luckman, S. (2018), 'Craft entrepreneurialism and sustainable scale: Resistance to and disavowal of the creative industries as champions of capitalist growth', *Cultural Trends,* 27(5): 313–326.
Luckman, S. and Andrew, J. (2020), *Craftspeople and Designer Makers in the Contemporary Creative Economy, Creative Working Lives,* Cham: Springer.
Oakdene Hollins and Dillon (2021), 'Socio-economic and environmental study of the Canadian remanufacturing sector and other value-retention processes in the context of a circular economy (No. En4- 438/2021E)'. *Environment and Climate Change Canada.*
Park, H.J. and Lin, L.M. (2020), 'Exploring attitude–behaviour gap in sustainable consumption: Comparison of recycled and upcycled fashion products', *Journal of Business Research,* 117: 623–628.
Pickvance, C. G. (2001), 'Four varieties of comparative analysis.' *Journal of Housing and the Built Environment,* 16(1), 7–28. Available at: http://www.jstor.org/stable/41107161.
Singh, J. and Ordoñez, I. (2016), 'Resource recovery from post-consumer waste: Important lessons for the upcoming circular economy', *Journal of Cleaner Production,* 134: 342–353.
Singh, J., Sung, K., Cooper, T., West, K. and Mont, O. (2019), 'Challenges and opportunities for scaling up upcycling businesses – The case of textile and wood upcycling businesses in the UK', *Resources, Conservation and Recycling,* 150: 104439.
Sung, K. and Abuzeinab, A. (2022), 'Challenges and Opportunities for Scaling Up Global Upcycling Towards Sustainable Production and Consumption', in W. Leal Filho, A.M. Azul, F. Doni, A.L. Salvia (eds), *Handbook of Sustainability Science in the Future: Policies, Technologies and Education by 2050,* 1–19, Cham: Springer.
Sung, K. and Cooper, T. (2015), 'Sarah Turner – Eco-artist and designer through craft-based upcycling', *Craft Research,* 6(1): 113–122.

Sung, K., Hughes, P. and Hsu, J. (2022), 'Exploratory multiple case study on successful upcycling businesses: ChopValue, Freitag and Pentatonic', *British Academy of Management (BAM) 2022 Conference*, Manchester.
Sung, K., Singh, J. and Bridgens, B. (2021), *State-of-the-Art Upcycling Research and Practice: Proceedings of the International Upcycling Symposium 2020*, Cham: Springer.
Van Hees, M. and Oskam, I. (2021), 'Unravelling Repurposing – A taxonomy for a promising circular business model strategy', *4th PLATE 2021 Virtual Conference*, Limerick.
Woolley, M. (2011), 'Beyond control: Rethinking industry and craft dynamics', *Craft Research*, 2(1): 11–36.
Yoo, F., Jung, H.J. and Oh, K.W. (2021), 'Motivators and barriers for buying intention of upcycled fashion products in China', *Sustainability*, 13(5): 2584.
Yu, S. and Lee, J. (2019), 'The effects of consumers' perceived values on intention to purchase upcycled products', *Sustainability*, 11(4): 1034.
Zhan, X. and Walker, S. (2019), 'Craft as leverage for sustainable design transformation: A theoretical foundation', *The Design Journal*, 22(4): 483–503.

Chapter 11

Key Challenges and Success Factors for Scaling Up Upcycling Businesses Across Industries and Countries

Kyungeun Sung, Amal Abuzeinab and Richie Moalosi

Introduction

In an age characterized by mounting environmental concerns, resource scarcity and a pressing need for sustainable solutions, upcycling has come to the forefront as a transformative concept that presents both challenges and opportunities for companies worldwide. Upcycling is the process of creatively reusing discarded materials to create products of higher value. It has gained momentum as a crucial component of the circular economy – an alternative to the current linear economy of take-make-use-dispose (Stahel, 2016). It employs a range of material processes and innovations to transform waste and used products, components, and materials into items of higher quality/value compared to their original compositional elements (Sung, 2017). Examples of upcycling comprise innovative or advanced forms of repair, reuse, repurposing, refurbishment, upgrade, redesign, remaking, remanufacturing, and recycling – all encompassed within the circular economy (Sung, 2023). Upcycling, as a principle, promotes greater material/resource efficiency and decreases solid waste, industrial energy consumption, and associated carbon emissions in manufacturing and processing with virgin materials, thus aiding the achievement of net zero (Caldera et al., 2022; Coppola et al., 2021). Upcycling not only reduces negative environmental impacts but also creates new job opportunities and reduces material costs (Jayasinghe et al., 2021; Pandit et al., 2020). Despite the advantages and the recent surge in the number of companies adopting upcycling, it remains a niche practice (Singh et al., 2019; Sung et al., 2019a). Expanding upcycling from a niche to a mainstream practice or scaling up has the potential to realize its benefits fully. Understanding the interconnected and mutually reinforcing challenges within the global upcycling value chain is pivotal to expanding global upcycling enterprises. For instance, the unfavourable perception of British consumers towards upcycled items impacts the market demand, thereby affecting the price and sales, resulting in a constrained market with limited high-quality product availability that, in turn, reinforces the consumers' negative view regarding upcycled goods (Singh et al., 2019). The interrelationships between factors make it difficult to establish which challenges can be prioritized in effectively scaling up upcycling businesses. In addition, existing studies of upcycling have been primarily focused on particular industry/sector (or waste streams) and countries/regions such as: (i) plastic waste upcycling (Chen et al., 2021; Blank et al., 2020; Lee et al., 2023; Zhao et al., 2022); (ii) fashion and textile upcycling (Aus et al., 2021; Marques et al., 2019; Haslinger et al., 2019); (iii) furniture upcycling (Kim & Yoon, 2016; Billinghurst, 2021; Baik & Kim, 2014); (iv) organic waste upcycling (Park et al., 2023; Areniello et al., 2023; Lasaridi & Stentiford, 2011); (v) paper upcycling (Ma et al., 2016; Zhou et al., 2021; Lopez & Faucheu, 2021); (vi) upcycling in the UK (Sung et al., 2019b; Singh et al., 2019; Sung & Cooper, 2015); (vii) upcycling in Europe (Sung et al., 2022; Ballerstedt et al., 2021; Brykała et al., 2022); (viii) upcycling in Americas (Bandara et al., 2020; Grasso et al., 2023; Jain & Gualandris, 2023); and (ix) upcycling in Asia (Kim, 2023; Yoo et al., 2021; Sharma & Gupta, 2014). The factors that influence the success of upcycling businesses across industries and countries are largely unknown. Therefore, there is a need for cross-industry and cross-country research that could potentially enable large-scale, global collective efforts to scale up upcycling to unlock opportunities for positive change. This study, therefore, aimed to identify critical global challenges and success factors for scaling up upcycling businesses across industries and countries.

Literature Review

Global Challenges for Scaling Up Upcycling Businesses

There are roughly eight common categories of cross-industry and cross-country challenges in upcycling: (i) design, making and production; (ii) materials; (iii) consumers; (iv) product; (v) marketing; (vi) space; (vii) regulations, standards and guarantees; and (viii) finance (Sung & Abuzeinab, 2021). Challenges in design, making and production are mainly related to limited knowledge, skills and capacities in upcycling businesses (Caldera et al., 2022; James & Kent, 2019; Singh et al., 2019), and time-consuming processes, especially for disassembly (or deconstruction) and manual work (handcraft) (Paras & Curteza, 2018; Chuma et al., 2019; Sung et al., 2017) with added complexities such as sorting and processing before manufacturing (Cumming, 2017). A lack of suitable production facilities and equipment has also been mentioned frequently by many scholars (Hjelmgren et al, 2014; Blank et al, 2020; Sung et al, 2020). There is particular difficulty in creating functional (e.g., furniture) or structural (e.g., construction) products by using upcycling (Ali et al, 2013; Oyenuga et al, 2017). There is also a risk of production for no demand (Hjelmgren et al., 2014). Challenges in materials are mostly related to material limitations (e.g., inconsistency and variety in type, size, colour, and patterns) that define processes (Ali et al., 2013; Cumming, 2017; Sung et al., 2017). There is difficulty in sourcing affordable, sufficient, various quality materials with certainty and stability (Han et al., 2015; James & Kent, 2019; Sung et al., 2020). Labour-intensive material sorting processes, storage of materials, and frequent transport associated with sourcing materials from multiple sources often result in high costs (Han et al., 2015; Sung et al., 2017; Paras & Curteza, 2018). Consumer challenges are largely negative perceptions and attitudes towards upcycled products: for example, some people think that upcycled products have poor finishing, reduced durability and low cost-effectiveness (Sung et al., 2017; Singh et al., 2019; Caldera et al., 2022). There is also a general lack of interest in, awareness or understanding of upcycling among consumers (Singh et al., 2019; Peirson-Smith & Evans, 2017). The main product challenge is limited, good or high-quality upcycled goods (Sung et al., 2017; Singh et al., 2019). The unique aesthetic of upcycled products is not necessarily to everyone's taste (Han et al., 2015), and there are potential health and safety risks from the often untested/unwarranted materials (Blank et al., 2020; Oyenuga et al., 2017). The main marketing challenges are high retail prices (Singh et al., 2019; Ozdamar-Ertekin, 2019) and difficult and expensive promotions (Caldera et al., 2022; Sung et al., 2020).

It is challenging to identify suitable marketplaces, buyers or consumers, and relevant sales and promotion opportunities offline and online, especially when/where there is no category for upcycled products (Sung et al., 2017). It is time-consuming and expensive to advertise and promote each one-off product, and it is not easy to get the narratives and keywords right for effective social media promotion (Singh et al., 2019). From the consumers' perspective, the current marketing of upcycled products does not offer adequate access to the products or a satisfactory shopping experience, such as difficulty finding individual shops/makers (Sung et al., 2020). The space issue is a general problem in which there is not enough reasonably priced room for storing materials or carrying out upcycling work, nor for the display of upcycled products for both upcycling businesses and retailers (Sung et al., 2017; Singh et al., 2019). Challenges in regulations, standards and guarantees include insufficient quality checks or warranties for upcycled goods, complicated waste management and donation laws/regulations, and inconsistent or non-uniform policies, regulations and standards for upcycling (Singh et al., 2019; Sung et al., 2020). Financial challenges are financial constraints or struggles that various upcycling businesses/initiatives have faced, as well as a lack of funding schemes and financial incentives (Caldera et al., 2022; Singh et al., 2019).

Global Success Factors for Scaling Up Upcycling Businesses

There are about ten common success factors identified for scaling up upcycling businesses worldwide. Firstly, it has been proposed that raising consumer/citizen awareness and education is necessary for scaling up. Pertaining to consumer education, the following recommendations have been made: (i) provision of practical workshops and design guides, (ii) involvement of citizens as volunteers and participants in upcycling events, initiatives and networks, (iii) awareness-raising campaigns and communication through various media, and (iv) environmental labelling of products (e.g., providing information regarding the environmental impact of both mass-produced and upcycled products) (Sung et al., 2017; Singh et al., 2019; Caldera et al., 2022). Second, suitable technologies, techniques and innovation in production would be helpful for scaling up. Various authors have stressed the importance of pinpointing well-suited methods and techniques, incorporating system-level, integrated strategies for waste treatment and production with upcycling, and exploring innovative approaches in new product development and technology solutions (Steinhilper & Hieber, 2001; Paras & Curteza, 2018; James & Kent, 2019). Third, targeting promising areas could be beneficial for effective scaling up, such as automotive products, electronics, furniture, clothing, patios, walkways, and planters (Steinhilper & Hieber, 2001; Oyenuga et al., 2017; Cumming, 2017; James & Kent, 2019). Fourth, various methods for enhancing material supply have been proposed for scaling up, including the implementation of material databases, the provision of materials and their testing, an integrated material sorting system within the current waste collection facilities, and autonomous resource hubs (James & Kent, 2019; Singh et al., 2019; Caldera et al., 2022; Sung, 2021). Fifth, several suggestions have been made regarding effective marketing for scaling up, including reasonable prices through Value Added Tax reduction, use of social media and relevant trade shows/fairs, and integration into the mainstream market, for example, by selling in high-street shops (Han et al., 2015; Sung et al., 2017; Caldera et al., 2022; Sung et al., 2020). Sixth, support is needed to better mobilize resources and knowledge for scaling up. Suggestions include training programmes for knowledge transfer and capacity building, a knowledge-sharing platform, creative business management support in general, a guild for workload sharing and human resource support and industrial equipment rental services (Sung et al., 2017; Singh et al., 2019; Caldera et al., 2022). Seventh, financial support and incentives for upcycling businesses and initiatives such as tax reduction, favourable loans, good business rates and funding schemes, have been suggested (Singh et al., 2019; Sung et al., 2020; Sung et al., 2019a). Eighth, it has been suggested that engaging and collaborating with different stakeholders would facilitate scaling up, such as more prominent brands, local communities, maker spaces, academic partners and local authorities (Han et al., 2015; Seravalli, 2016; Caldera et al., 2022). Ninth, legal support has been recommended in relation to the donation and collection of waste materials, as well as regulatory considerations in the transfer of new technologies to commercial markets (Blank et al., 2020; Sung et al., 2020). Tenth, appropriate business space has been suggested for material storage, deconstruction of materials, and new product development (Sung et al., 2017; James & Kent, 2019).

Methodology

This study aimed to identify the critical global challenges and success factors for scaling up upcycling businesses across industries and countries. Interpretive

Structural Modelling (ISM) was employed as a qualitative and interpretive method used to structure various factors into a hierarchy based on their importance and visualize the interaction between these factors to address these factors in order of importance (Attri et al., 2013; Abuzeinab et al., 2017). Two exploratory ISM workshops and one confirmative ISM workshop were carried out between July 2022 and June 2023.

Interpretive Structural Modelling (ISM)

Interpretive Structural Modelling (ISM) is a recognized methodology for identifying the relationships among individual elements that define a problem, originated as an interactive group learning process (Attri et al., 2013). ISM has been widely used to explore various challenges/barriers/obstacles and success factors in different areas and disciplines. The process involved in the ISM method includes several steps. Step 1 is to identify variables affecting the system under examination. Step 2 is to create a structural self-interaction matrix (SSIM) to display dependence among all possible pairs of elements, determined by a contextual relationship showing which elements affect others. Step 3 involves creating a reachability matrix (RM) from the SSIM and verifying its transitivity. The transitivity is a basic assumption in ISM. This means that if variable X is related to Y and Y is linked to Z, then X must be related to Z as well. Step 4 involves categorizing variables based on their driving and dependence power using MICMAC (Matriced' Impacts Croise's Multiplication Applique and Classment) analysis – or Cross-Impact Matrix Multiplication Applied to Classification. Step 5 is to partition the reachability matrix into distinct levels. Step 6 entails drawing a directed graph based on the relationships present in the reachability matrix while removing the transitive links – in other words, developing the ISM model. Step 7 is to check for conceptual inconsistency and make any necessary modifications (Attri et al., 2013; Ravi & Shankar, 2005; Shahabadkar et al., 2012).

Interpretive Structural Modelling (ISM) Workshops

Two online ISM workshops of an exploratory nature were held in September 2022, followed by a third and final ISM workshop in June 2023, which took place in person to verify and validate the findings. Nine upcycling experts from academia and industry in seven countries across Africa, Asia and Europe (Botswana, India, Kenya, Nigeria, South Africa, South Korea, UK) participated in the exploratory online workshops. The in-person confirmative workshop involved seven global experts in upcycling and sustainable production and consumption from Germany, Japan, Netherlands, Sweden and the UK. At the beginning of each workshop, the aim and objectives of the workshop and procedures were explained. The workshop was structured with each challenge or success factor presented alongside others (Figure 12.1). Participants were asked as to whether the presented factor would impact the other factors. Individual perspectives were gathered and group consensus was achieved, documenting key points and conclusions through note-taking and video recording of the workshops with participants' approval.

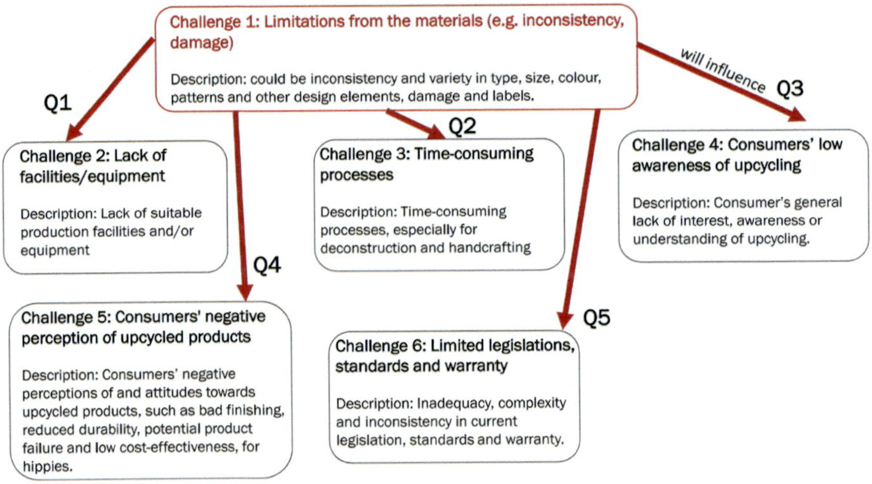

Figure 11.1 Illustrative diagram example presented during the ISM workshops.

Development of ISM Models

Identifying Key Variables

The factors (global challenges and success factors identified from the literature review) were narrowed down to six each (the maximum, manageable number) for the ISM workshops. To narrow down the factors, an expert online survey was conducted between July and August 2022. This was sent to 21 global upcycling experts via email and 14 responses were collected (67% response rate). The respondents were nine academics and five practitioners (nine men and five women) with expertise in upcycling from 12 countries on five continents (Australia, Botswana, Chile, Costa Rica, Ghana, Hong Kong, India, Nigeria, South Korea, Sweden, UK, Zimbabwe). The survey asked the respondents to 'select up to six key global challenges for scaling up upcycling businesses' and 'select up to six key success factors for scaling up upcycling businesses' from the listed factors based on the literature review. The online survey results showed that the majority of the respondents agreed (50% and above) with six common global challenges and success factors for scaling up upcycling businesses as follows: limitations from the materials (Challenge 1), lack of facilities/equipment (C2), time-consuming processes (C3), consumers' low awareness of upcycling (C4), consumers' negative perception of upcycled products (C5) and limited legislation, standards, and warranty (C6) as global common challenges; and improved material provision (Opportunity 1), targeting high-opportunity product categories (O2), suitable technologies, techniques, and innovation in production (O3), varied stakeholder involvement (O4), involving citizens in upcycling initiatives/events (O5), and awareness-raising activities (O6) as global common success factors.

Developing SSIMs for Global Challenges and Success Factors

The responses and consensus of the workshop participants were analysed to establish pair-wise relationships using the Structural Self-Interaction Matrix (SSIM). Following this analysis, a contextual relationship of influence was selected, and four symbols were used to indicate the direction of the relationship between two given factors (i and j):

- V: Challenge/success factor i will influence challenge/success factor j, but not in both directions.
- A: Challenge/success factors j will influence challenge/success factor i, but not in both directions.
- X: Challenge/success factors i and j will influence each other, and
- O: Challenge/success factors i and j are unrelated.

According to the SSIM for global challenges for scaling up upcycling businesses (Table 12.1), the confirmative workshop mostly validates the results of the exploratory workshops. However, there were two controversial factors: lack of facilities/equipment and time-consuming processes. In general, most factors seem to have interdependent relationships and influences on one another. The SSIM for global success factors for scaling up upcycling businesses (Table 12.2) shows that the confirmative workshop validates the results of the exploratory workshops without any contentious factors. Most factors exhibit interdependent relationships.

Developing RMs from SSIMs

The RM (Reachability Matrix) was obtained by converting the SSIM into a binary matrix. V, A, X and O were substituted by 1 or 0 depending on the case. The substitution rules were as follows:

- if the (i, j) entry in the SSIM is V, then the (i, j) entry in the reachability matrix becomes 1, and the (j, i) entry becomes 0.

Table 11.1 SSIM for global challenges

	Exploratory/Confirmative					
	C1	**C2**	**C3**	**C4**	**C5**	**C6**
C1. Limitations from the materials		X/A	X/V	A/A	A/X	X/X
C2. Lack of facilities/equipment			X/V	A/O	X/O	X/A
C3. Time-consuming processes				X/O	V/V	X/O
C4. Consumers' low awareness of upcycling					X/V	X/X
C5. Consumers' negative perception of upcycled products						X/X
C6. Limited legislation, standards, and warranty						

Table 11.2 SSIM for global success factors

	Exploratory/Confirmative					
	O1	**O2**	**O3**	**O4**	**O5**	**O6**
O1. Improved material provision		V/X	X/X	X/X	X/X	X/X
O2. Targeting high-opportunity product categories			X/X	X/X	X/X	A/X
O3. Suitable technologies, techniques, and innovation in production				X/X	A/X	A/X
O4. Varied stakeholder involvement					X/X	X/X
O5. Involving citizens in upcycling initiatives/events						X/X
O6. Awareness-raising activities						

- if the (i, j) entry in the SSIM is A, then the (i, j) entry in the reachability matrix becomes 0, and the (j, i) entry becomes 1.
- if the (i, j) entry in the SSIM is X, then the (i, j) entry in the reachability matrix becomes 1, and the (j, i) entry also becomes 1, and
- if the (i, j) entry in the SSIM is O, then the (i, j) entry in the reachability matrix becomes 0, and the (j, i) entry also becomes 0.

Following these rules, the RMs for global challenges and success factors for scaling up upcycling businesses are shown in Tables 12.3–12.4. As transitivity is absent, the RM matrix and final RM coincide. Tables 12.3–12.4 show the calculation of driving and dependence power within the framework of the ISM model. The driving power of each factor is the total number of factors, including itself, that it may influence. Dependence power is the total number of factors that may help address it. Based on the driving and dependence power, four clusters can classify

Table 11.3 RM for global challenges

	Exploratory/Confirmative						
	C1	**C2**	**C3**	**C4**	**C5**	**C6**	**Driving power**
C1. Limitations from the materials	1/1	1/0	1/1	0/0	0/1	1/1	4/4
C2. Lack of facilities/equipment	1/1	1/1	1/1	0/0	1/0	1/1	5/4
C3. Time-consuming processes	1/1	1/1	1/1	1/0	1/1	1/0	6/4
C4. Consumers' low awareness of upcycling	1/1	1/0	1/0	1/1	1/1	1/1	6/4
C5. Consumers' negative perception of upcycled products	1/1	1/0	0/1	1/0	1/1	1/1	5/4
C6. Limited legislation, standards, and warranty	1/1	1/1	1/0	1/1	1/1	1/1	6/5
Dependence	6/6	6/3	5/4	4/2	5/5	6/5	32/25

Table 11.4 RM for global success factors

	Exploratory/Confirmative						
	O1	**O2**	**O3**	**O4**	**O5**	**O6**	**Driving power**
O1. Improved material provision	1/1	1/1	1/1	1/1	1/1	1/1	6/6
O2. Targeting high-opportunity product categories	0/1	1/1	1/1	1/1	1/1	0/1	4/6
O3. Suitable technologies, techniques, and innovation in production	1/1	1/1	1/1	1/1	0/1	0/1	4/6
O4. Varied stakeholder involvement	1/1	1/1	1/1	1/1	1/1	1/1	6/6
O5. Involving citizens in upcycling initiatives/events	1/1	1/1	1/1	1/1	1/1	1/1	6/6
O6. Awareness-raising activities	1/1	1/1	1/1	1/1	1/1	1/1	6/6
Dependence	5/6	6/6	6/6	6/6	5/6	4/6	32/36

global challenges and success factors: autonomous, dependent, linkage, and independent/driver factors. The subsequent section provides a more detailed explanation of this classification and its limitations.

Classifying Global Challenges and Success Factors – MICMAC Analysis

Based on the driving power and dependence power generated in Tables 12.3–12.4, MICMAC analysis classified global challenges and success factors for scaling up upcycling businesses, as shown in Figures 12.2–12.3. The diagrams for global challenges (Figure 12.2) indicate that no challenge comes under an autonomous

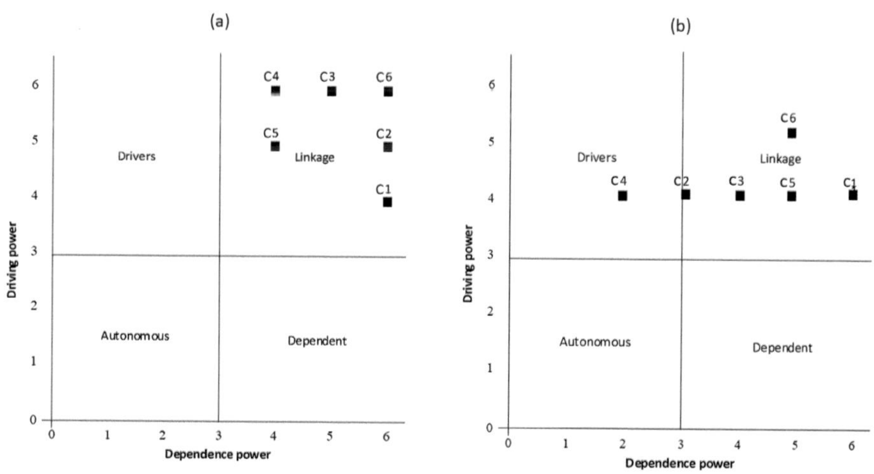

Figure 11.2 MICMAC analysis for global challenges: (a) exploratory workshops; (b) confirmative workshop.

cluster. Autonomous challenges are typically weak drivers, with little dependence and relative disconnection from the system. These challenges have minimal impact on other challenges of the system. There are also no dependent challenges. Dependent challenges require the resolution and removal of other underlying challenges. There might be one challenge within the driver cluster (not identified during exploratory workshops but acknowledged in the confirmative workshop) – C4. 'Consumers' low awareness of upcycling'. Driver challenges, which possess strong driving power but weak dependence power, should be prioritized and addressed first since they can impact all other challenges. Except for C4, all other challenges belong to the linkage cluster, which has a strong driving power and strong dependence. These challenges are unstable due to their sensitivity to external factors, and any action on them will have a feedback effect on them.

Figure 12.3 illustrates the global success factors diagrams, which indicate the absence of autonomous, dependent or driver factors. All success factors are contained in the linkage cluster, displaying strong driving power and strong dependence. Each factor's score and position on the diagram differ between the exploratory and the confirmative workshops, with higher scores and tighter relationships between factors from the confirmative workshop. Nonetheless, the linkage cluster remains the same, thus affirming and validating the results.

Partitioning the RMs into Different Levels

From the final RM, the reachability and antecedent set for each factor were derived, after which the intersection of these sets was identified. Tables 12.5–12.6 show RMs in terms of global challenges and success factors for scaling up upcycling businesses. The 1st level was assigned to the challenges/success factors where the reachability and intersection sets were identical in the first iteration. The

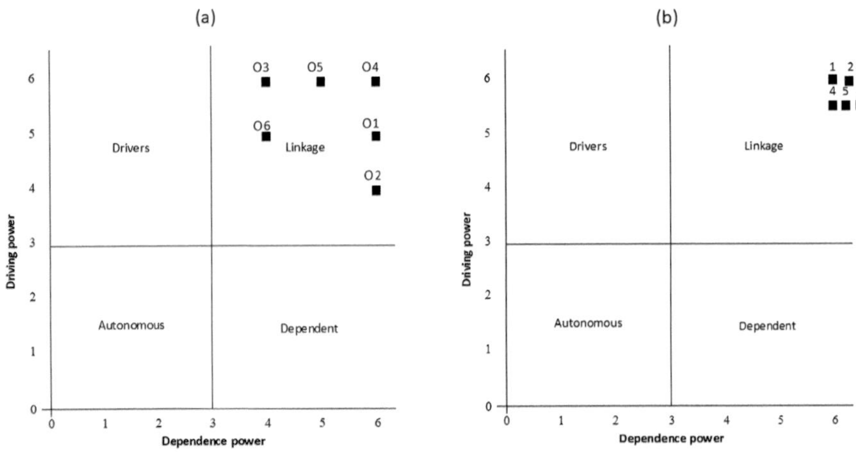

Figure 11.3 MICMAC analysis for global success factors: (a) exploratory workshops; (b) confirmative workshop.

levels of all the challenges/success factors were established by duplicating this procedure. After determining a factor's level, it was removed from the list of remaining challenges/success factors for the subsequent iteration. The levels denote the priorities in the ISM models' hierarchy, with the first level being the least important.

Table 11.5 RM into levels for global challenges

Challenge		**Reachability set**	**Antecedent set intersect**	**Intersection set**	**Level**
Iteration 1					
C1. Limitations from the materials	Ex*	C1,2,3,6	C1,2,3,4,5,6	C1,2,3,6	1st
	Con	C1,3,5,6	C1,2,3,4,5,6	C1,3,5,6	1st
C2. Lack of facilities/ equipment	Ex	C2,1,3,5,6	C2,1,3,4,5,6	C2,1,3,5,6	1st
	Con	C2,1,3,6	C2,3,6	C2,3,6	
C3. Time-consuming processes	Ex	C3,1,2,4,5,6	C3,1,2,4,6	C3,1,2,4,6	
	Con	C3,1,2,5	C3,1,2,5	C3,1,2,5	1st
C4. Consumers' low awareness of upcycling	Ex	C4,1,2,3,5,6	C4,3,5,6	C4,3,5,6	
	Con	C4,1,5,6	C4,6	C4,6	
C5. Consumers' negative perception of upcycled products	Ex	C5,1,2,4,6	C5,2,3,4,6	C5,2,4,6	
	Con	C5,1,3,6	C5,1,3,4,6	C5,1,3,6	1st
C6. Limited legislation, standards, and warranty	Ex	C6,1,2,3,4,5	C6,1,2,3,4,5	C6,1,2,3,4,5	1st
	Con	C6,1,2,4,5	C6,1,2,4,5	C6,1,2,4,5	1st
Iteration 2					
C2. Lack of facilities/ equipment	Con	C2	C2	C2	2nd
C3. Time-consuming processes	Ex	C3,4,5	C3,4	C3,4	
C4. Consumers' low awareness of upcycling	Ex	C4,1,5	C4,3,5	C4,5	
	Con	C4	C4	C4	2nd
C5. Consumers' negative perception of upcycled products	Ex	C5,4	C5,3,4	C5,4	2nd
Iteration 3					
C3. Time-consuming processes	Ex	C3,4	C3,4	C3,4	3rd
C4. Consumers' low awareness of upcycling	Ex	C4,1	C4,3	C4	

* Ex is exploratory workshop results; Con is confirmative workshop results

Table 11.6 RM into levels for global success factors

Challenge		Reachability set	Antecedent set intersect	Intersection set	Level
Iteration 1					
O1. Improved material provision	Ex	O1,2,3,4,5,6	O1,3,4,5,6	O1,3,4,5,6	
	Con	O1,2,3,4,5,6	O1,2,3,4,5,6	O1,2,3,4,5,6	1st
O2. Targeting high-opportunity product categories	Ex	O2,3,4,5	O2,1,3,4,5,6	O2,3,4,5	1st
	Con	O2,1,3,4,5,6	O2,1,3, 4,5,6	O2,1,3,4,5,6	1st
O3. Suitable technologies, techniques, and innovation in production	Ex	O3,1,2,4	O3,1,2,4,5,6	O3,1,2,4	1st
	Con	O3,1,2,4,5,6	O3,1,2,4,5,6	O3,1,2,4,5,6	1st
O4. Varied stakeholder involvement	Ex	O4,1,2,3,5,6	O4,1,2,3,5,6	O4,1,2,3,5,6	1st
	Con	O4,1,2,3,5,6	O4,1,2,3,5,6	O4,1,2,3,5,6	1st
O5. Involving citizens in upcycling initiatives/ events	Ex	O5,1,2,3,4,6	O5,2,4,6	O5,2,4	
	Con	O5,1,2,3,4,6	O5,1,2,3,4,6	O5,1,2,3,4,6	1st
O6. Awareness-raising activities	Ex	O6,1,2,3,4,5	O6,1,4,5	O6,1,4,5	
	Con	O6,1,2,3,4,5	O6,1,2,3,4,5	O6,1,2,3,4,5	1st
Iteration 2					
O1. Improved material provision	Ex	O1,5,6	O1,5,6	O1,5,6	2nd
O5. Involving citizens in upcycling initiatives/ events	Ex	O5,1,6	O5,6	O5,6	3rd
O6. Awareness-raising activities	Ex	O6,1,5	O6,1,5	O6,1,5	2nd

Developing the ISM Models for Global Challenges and Success Factors

From the analyses, including the levels determined, the ISM models were developed for global challenges and success factors (Figures 12.4–12.5). The ISM model for critical global challenges from the exploratory workshops shows four levels, whereas the one from the confirmative workshop shows two levels (Figure 12.4). More levels indicate stronger hierarchical relationships between the factors, whereas fewer levels indicate less hierarchical (and more equal) relationships. The first ISM model (Figure 12.4 left) shows that all the challenges are interlinked – arrow directions denoting the relationships. The model was structured on four levels. At the base of the structure, 'consumers' low awareness of upcycling' will need to be addressed before moving up to Level 3 (C3. Time-consuming processes) and then Level 2 (C5. Consumers' negative perception of upcycled products). Three challenges at the top of the structure (C1. Limitations from the materials, C2. Lack of facilities/equipment, and C6. Limited legislation, standards, and warranty) can be addressed at the end. The second ISM model (Figure 12.4 right) confirms that all the challenges are interlinked, but the model was structured on two levels only. The model confirms that C4. 'Consumers' low awareness of upcycling' is the most critical, high-priority challenge that needs to be addressed first, and that is C1. 'Limitations from the

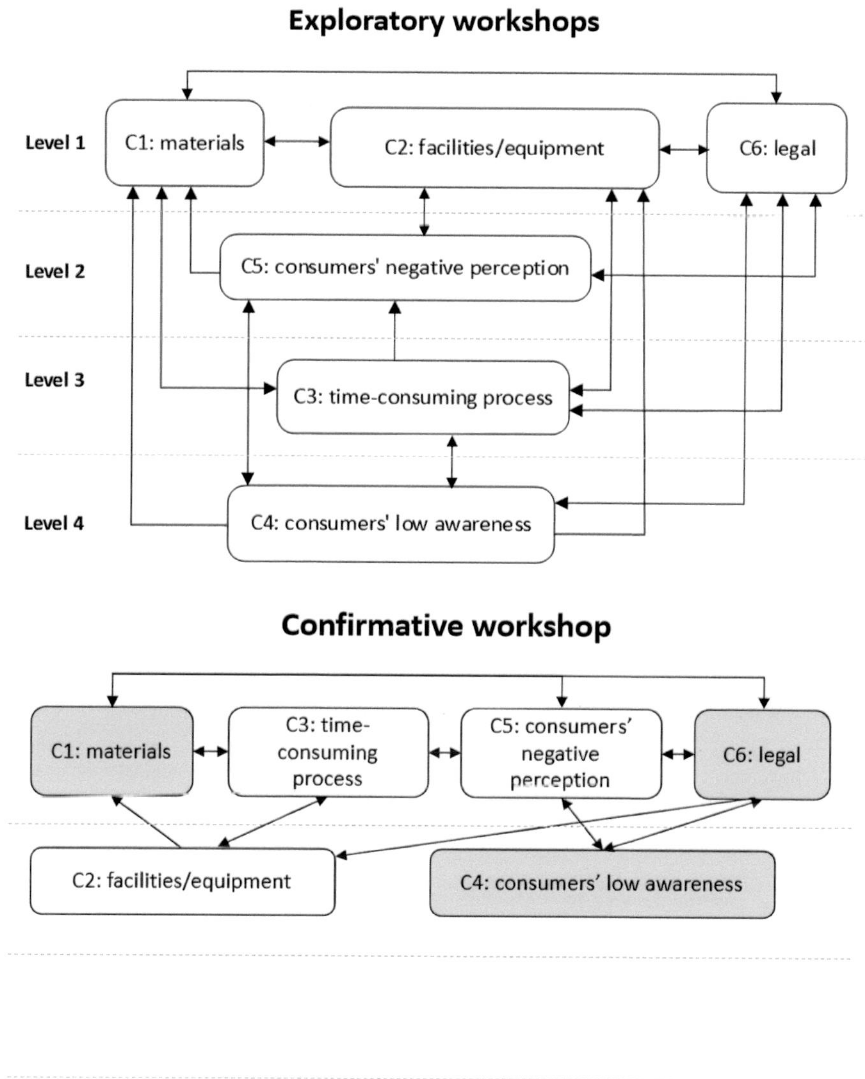

Figure 11.4 ISM models for critical global challenges (grey cells confirmed).

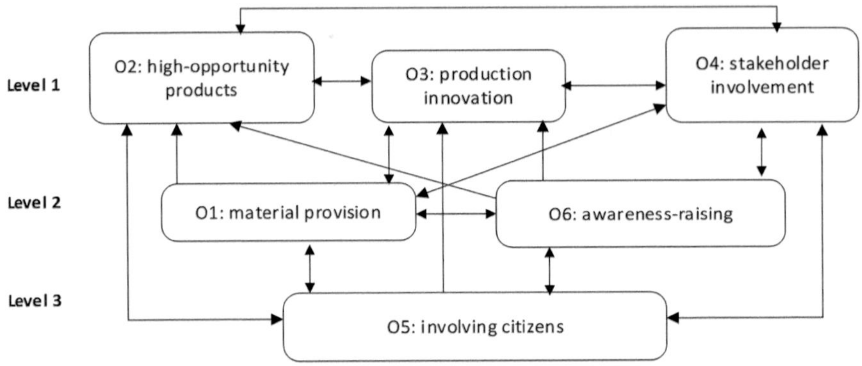

Figure 11.5 ISM model for critical global success factors from the exploratory workshops.

materials' and C6. 'Limited legislation, standards, and warranty' are the least essential challenges that could be addressed later.

The ISM model for key global success factors from the exploratory workshops (Figure 12.5) shows that all opportunities are interlinked. The model was structured on three levels. At the base of the structure, priority should be given to O5. 'Involving citizens for upcycling initiatives/events' before progressing to Level 2, which encompasses O1. 'Improved material provision' and O6. 'Awareness-raising activities.' Subsequently, the focus can shift to Level 1, which includes O2. 'Targeting high-opportunity product categories', O3. 'Suitable technologies, techniques, and innovation in production', and O4. 'Varied stakeholder involvement'. The confirmative workshop, however, classified all the success factors on one level, creating an ISM model structured on one level. This means that six global success factors are equally essential and can be addressed simultaneously. Therefore, it is not meaningful to visualize the model in this case.

Discussion and Conclusion

Overall, the results showed that all critical global challenges and opportunities are interlinked (within the category) and that crucial global success factors reflect critical global challenges. For example, the most critical success factor turned out to be 'involving citizens in upcycling initiatives/events' when the most critical challenge appeared to be 'consumers' low awareness of upcycling'. From the results, the following practical implications could be derived. The priority action as an intervention for scaling up global upcycling businesses should be 'involving citizens in upcycling initiatives and events for awareness raising including campaign, communication, training, and education' (O5 and O6) to address 'consumers' low awareness of upcycling' (C4) and 'consumers' negative perception of upcycled products' (C5). The second priority interventions could be 'improving material provision' (O1) and addressing 'time-consuming processes' (C3) and a lack of facilities and equipment (C2) by 'suitable technologies, techniques, and innovation in production' (O3).

There were some differences in the ISM models between the exploratory and the confirmative workshops regarding the global challenges. The differing demographics of the study participants may account for these results. Specifically, more participants from developing countries in Africa and Asia participated in the exploratory workshops. In contrast, all participants in the confirmative workshop were from developed countries in Asia and Europe. People in developing countries may be inclined to value mass production and harbour negative perceptions towards using waste materials due to poverty connotations instead of environmentalism (Sung, 2015). Hence, consumers' unfavourable views towards upcycled products could carry more significance in developing countries than in developed countries. People in developed countries, however, may have encountered numerous luxury sustainable goods made through upcycling and various other handcrafted goods using time-consuming processes (Phau et al., 2022; Keith & Silies, 2015). Therefore, they may not necessarily perceive time-consuming processes or consumers' negative perception of upcycled products as significant challenges. At the same time, they may also have greater confidence in technology or its implementation and believe that the main obstacle or challenge affecting material supply and production processes is the current shortage of facilities and equipment. Similarly, concerning global success factors, the study participants in developing countries seemed to have a stronger urge to enhance their citizens' awareness of upcycling and improve material provision. Again, this could be attributed to the poverty connotations and less efficient waste management systems and other infrastructure that utilize waste and used materials, components and products in developing countries (Mmereki et al., 2016).

We acknowledge the limitation of this study as our approach bundled data from all countries, which fails to reflect the differences of various contexts, as demonstrated by the slight discrepancies found in the ISM models. The differences may be dependent on the socioeconomic and cultural backgrounds of the participants. Our approach also precludes comparative analysis between countries, continents, or industries/sectors. The primary objective of this study, however, was to determine the key global challenges and success factors, irrespective of contextual variations. Our main contribution to knowledge is, therefore, the identification of universal factors that can serve as a starting point for designing and developing effective interventions to scale up upcycling businesses on a global level.

Acknowledgements

This research was funded by the AHRC Research Networking grant for the International Upcycling Research Network, grant number AH/W007134/1. The generous Research Award from DMU supported the work. We thank all the study participants. We thank Nicky Harding, Stuart Lawson, Kate Cheyne, Kelley Wilder,

Siobhan Keenan, Christine White, Jill Cowley, Deborah Cartmell, Mike Kagioglou, and Katie Normington for supporting this research at DMU.

References

Abuzeinab, A., Arif, M. and Qadri, M.A. (2017), 'Barriers to MNEs green business models in the UK construction sector: An ISM analysis', *Journal of Cleaner Production*, 160: 27–37.

Ali, N.S., Khairuddin, N.F. and Abidin, S.Z. (2013), 'Upcycling: re-use and recreate functional interior space using waste materials', *Proceedings of the 15th International Conference on Engineering and Product Design Education*, 798–803, Dublin.

Areniello, M., Matassa, S., Esposito, G. and Lens, P.N. (2023), 'Biowaste upcycling into second-generation microbial protein through mixed-culture fermentation', *Trends in Biotechnology*, 41(2): 197–213.

Attri, R., Dev, N. and Sharma, V. (2013), 'Interpretive structural modelling (ISM) approach: An overview', *Research Journal of Management Sciences*, 2319(2): 1171.

Aus, R., Moora, H., Vihma, M., Unt, R., Kiisa, M. and Kapur, S. (2021), 'Designing for circular fashion: Integrating upcycling into conventional garment manufacturing processes', *Fashion and Textiles*, 8: 1–18.

Baik, E. and Kim, J. (2014), 'A study on the trend analysis of upcycling furniture design', *Journal of the Korea Furniture Society*, 25(2): 112–121.

Ballerstedt, H., Tiso, T., Wierckx, N., Wei, R., Averous, L., Bornscheuer, U., O'Connor, K., Floehr, T., Jupke, A. and Klankermayer, J. (2021), 'MIXed plastics biodegradation and UPcycling using microbial communities: EU Horizon 2020 project MIX-UP started January 2020', *Environmental Sciences Europe*, 33(1): 99.

Bandara, N., Hettiarachchi, H., Jensen, E. and Binoy, T.H. (2020), 'Upcycling potential of industrial waste in soil stabilization: use of kiln dust and fly ash to improve weak pavement subgrades encountered in Michigan, USA', *Sustainability*, 12(17): 7226.

Billinghurst, C. (2021), 'Professional Upcycling in Partnership with Sustainability', in K. Sung, J. Singh, B. Bridgens (eds), *State-of-the-Art Upcycling Research and Practice*, 111–117, Cham: Springer.

Blank, L.M., Narancic, T., Mampel, J., Tiso, T. and O'Connor, K. (2020), 'Biotechnological upcycling of plastic waste and other non-conventional feedstocks in a circular economy', *Current Opinion in Biotechnology*, 62: 212–219.

Brykała, D., Pogodziński, P.M. and Piotrowski, R. (2022), 'Traces of disappearing heritage: Upcycling of wooden vessels preserved in the vernacular architecture of a large river valley in Central Europe', *Rural History*, 34(2): 243–261.

Caldera, S., Jayasinghe, R., Desha, C., Dawes, L. and Ferguson, S. (2022), 'Evaluating barriers, enablers and opportunities for closing the loop through 'waste upcycling': A systematic literature review', *Journal of Sustainable Development of Energy, Water and Environment Systems*, 10(1): 1–20.

Chen, X., Wang, Y. and Zhang, L. (2021), 'Recent progress in the chemical upcycling of plastic wastes', *ChemSusChem*, 14(19): 4137–4151.

Chuma, C., Muza, T. and Chipambwa, W. (2019), 'An examination of the upcycling design process of used clothes among fashion design students: Towards an approach for sustainable fashion design practice', *International Journal of Costume and Fashion*, 19(2): 59–69.

Coppola, C., Vollero, A. and Siano, A. (2021), 'Consumer upcycling as emancipated self-production: Understanding motivations and identifying upcycler types', *Journal of Cleaner Production*, 285: 124812.

Cumming, D. (2017), 'A Case Study Engaging Design for Textile Upcycling', *Journal of Textile Design Research and Practice*, 4(2): 113–128.

Grasso, S., Fu, R., Goodman-Smith, F., Lalor, F. and Crofton, E. (2023), 'Consumer attitudes to upcycled foods in US and China', *Journal of Cleaner Production*, 388: 135919.

Han, S., Tyler, D. and Apeagyei, P. (2015), 'Upcycling as a design strategy for product lifetime optimisation and societal change', *Proceedings of the PLATE Conference 2015*, Nottingham: Nottingham Trent University.

Haslinger, S., Hummel, M., Anghelescu-Hakala, A., Määttänen, M. and Sixta, H. (2019), 'Upcycling of cotton polyester blended textile waste to new man-made cellulose fibers', *Waste Management*, 97: 88–96.

Hjelmgren, D., Salomonson, N. and Ekström, K.M. (2014), 'Upcycling of Pre-Consumer Waste: Opportunities and Barriers in the Furniture and Clothing Industries', in K. Ekström (ed), *Waste Management and Sustainable Consumption*, 187–198, Abingdon: Routledge.

Jain, S. and Gualandris, J. (2023), 'When does upcycling mitigate climate change? The case of wet spent grains and fruit and vegetable residues in Canada', *Journal of Industrial Ecology*, 27(2): 522–534.

James, A.S.J. and Kent, A. (2019), 'Clothing Sustainability and Upcycling in Ghana', *Fashion Practice*, 11(3): 375–396.

Jayasinghe, R., Liyanage, N. and Baillie, C. (2021), 'Sustainable waste management through eco-entrepreneurship: An empirical study of waste upcycling eco-enterprises in Sri Lanka', *Journal of Material Cycles and Waste Management*, 23: 557–565.

Keith, S. and Silies, M. (2015), 'New life luxury: Upcycled Scottish heritage textiles', *International Journal of Retail & Distribution Management*, 43(10/11): 1051–1064.

Kim, J. and Yoon, Y. (2016), 'A study on the up-cycling furniture design using the Korea traditional patchwork technique-focusing on the production of stitch furniture Design', *Journal of the Korea Furniture Society*, 27(1): 49–58.

Kim, S. (2023), 'Review of food upcycling in South Korea: regulation, limitation, and prospects', *Food Science and Biotechnology*, 32(1): 1–10.

Lasaridi, K. and Stentiford, E. (2011), 'Upcycling organic waste in a world of thinly distributed resources', *Waste Management & Research*, 29(11): 1115–1116.

Lee, S., Lee, Y.R., Kim, S.J., Lee, J. and Min, K. (2023), 'Recent advances and challenges in the biotechnological upcycling of plastic wastes for constructing a circular bioeconomy', *Chemical Engineering Journal*, 454: 140470.

Lopez, N.S.R. and Faucheu, J. (2021), 'Exploring local recirculation of paper waste through upcycling and artistic recycling', *Proceedings of the Design Society*, 1481–1490.

Ma, Y., Hummel, M., Määttänen, M., Särkilahti, A., Harlin, A. and Sixta, H. (2016), 'Upcycling of waste paper and cardboard to textiles', *Green Chemistry*, 18(3): 858–866.

Marques, A.D., Moreira, B., Cunha, J. and Moreira, S. (2019), 'From waste to fashion – a fashion upcycling contest', *Procedia CIRP*, 84: 1063–1068.

Mmereki, D., Baldwin, A. and Li, B. (2016), 'A comparative analysis of solid waste management in developed, developing and lesser developed countries', *Environmental Technology Reviews*, 5(1): 120–141.

Oyenuga, A.A. and Bhamidimarri, R. (2017), 'Upcycling ideas for sustainable construction and demolition waste management: Challenges, opportunities and boundaries',

International Journal of Innovative Research in Science, Engineering and Technology, 6(3): 4066–4079.

Ozdamar-Ertekin, Z. (2019), 'Can luxury fashion provide a roadmap for sustainability?', *Markets, Globalization & Development Review*, 4(1): 1–21.

Pandit, P., Singha, K., Kumar, L., Shrivastava, S. and Yashraj, V. (2020), 'Business paradigm shifting: Opportunities in the 21st century on fashion from recycling and upcycling', in P. Pandit, S. Ahmed, K. Singha, S. Shrivastava (eds), *Recycling from waste in fashion and textiles: A sustainable and circular economic approach*, 151–176, Beverly: Scrivener Publishing LLC.

Paras, M.K. and Curteza, A. (2018), 'Revisiting upcycling phenomena: a concept in clothing industry', *Research Journal of Textile and Apparel*, 22(1): 46–58.

Park, H., Joo, J., Kim, J., Lee, J. and Kim, S. (2023), 'A sustainable strategy for organic waste upcycling: Concurrent production of energy and Li-ion battery anode from chicken litter', *Energy*, 278: 127797.

Peirson-Smith, A. and Evans, S. (2017), 'Fashioning green words and eco language: an examination of the user perception gap for fashion brands promoting sustainable practices', *Fashion Practice*, 9(3): 373–397.

Phau, I., Akintimehin, O. and Lee, S. (2022), 'Investigating consumers' brand desirability of "upcycled" luxury: the many faces of designer facemasks', *Journal of Fashion Marketing and Management: An International Journal*, Vol. ahead-of-print (No. ahead-of-print).

Ravi, V. and Shankar, R. (2005), 'Analysis of interactions among the barriers of reverse logistics', *Technological Forecasting and Social Change*, 72(8): 1011–1029.

Seravalli, A. (2016), 'Making an upcycling station: makers' culture, cross-sector collaborations and citizens' participation for new services and practices within waste handling', Available at: https://www.diva-portal.org/smash/get/diva2:1410357/FULLTEXT01.pdf

Shahabadkar, P., Hebbal, S.S. and Prashant, S. (2012), 'Deployment of Interpretive Structural Modeling methodology in supply chain management – An overview', *International Journal of Industrial Engineering and Production Research*, 23(3): 195–205.

Sharma, K. and Gupta, A. (2014), 'Scope of up-cycling in India' in N.J. Rajaram, G.H.S. Prasad, S.K.M. Belli (eds), *Contemporary Issues and Trends in Fashion, Retail and Management*, 383–387, Hyderabad: National Institute of Fashion Technology.

Singh, J., Sung, K., Cooper, T., West, K. and Mont, O. (2019), 'Challenges and opportunities for scaling up upcycling businesses – The case of textile and wood upcycling businesses in the UK', *Resources, Conservation and Recycling*, 150: 104439.

Stahel, W.R. (2016), 'The circular economy', *Nature*, 531(7595): 435–438.

Steinhilper, R. and Hieber, M. (2001), 'Remanufacturing-the key solution for transforming "downcycling" into "upcycling" of electronics', *IEEE*, 161–166.

Sung, K. (2015), 'A review on upcycling: Current body of literature, knowledge gaps and a way forward', *International Conference on Environmental, Cultural, Economic and Social Sustainability*, 17(4): 28-40.

Sung, K. (2017), *Sustainable production and consumption by upcycling: Understanding and scaling up niche environmentally significant behaviour*, Nottingham: Nottingham Trent University.

Sung, K. (2021), 'Evaluating two interventions for scaling up upcycling: Community event and upcycling plaza', *IOP Conference Series: Materials Science and Engineering*, 1196(1): 012001.

Sung, K. (2023), 'Understanding upcycling and circular economy and their interrelationships through literature review for design education', *Proceedings of the Design Society*, 3721–3730.

Sung, K. and Abuzeinab, A. (2021), 'Challenges and opportunities for scaling up global upcycling towards sustainable production and consumption: Literature review'. in W. Leal Filho, A.M. Azul, F. Doni, A.L. Salvia (eds), *Handbook of Sustainability Science in the Future: Policies, Technologies and Education by 2050*, 1–19, Cham: Springer.
Sung, K. and Cooper, T. (2015) 'Sarah Turner–Eco-artist and designer through craft-based upcycling', *Craft Research*, 6(1): 113–122.
Sung, K., Cooper, T. and Kettley, S. (2019a), 'Developing interventions for scaling up uk upcycling', *Energies*, 12(14): 2778.
Sung, K., Cooper, T. and Kettley, S. (2019b), 'Factors Influencing Upcycling for UK Makers', *Sustainability*, 11(3): 870.
Sung, K., Ku, L. and Yoon, J. (2022), 'Predictors of upcycling in the highly-industrialised West: A survey across three continents of Australia, Europe, and North America', *Sustainability*, 15(2): 1461.
Sung, K., Cooper, T., Ramanathan, U. and Singh, J. (2017), 'Challenges and support for scaling up upcycling businesses in the UK: Insights from small-business entrepreneurs', *Proceedings of the PLATE Conference 2017*, 397–401, Amsterdam: IOS Press.
Sung, K., Cooper, T., Oehlmann, J., Singh, J. and Mont, O. (2020), 'Multi-stakeholder perspectives on scaling up UK fashion upcycling businesses', *Fashion Practice*, 12(3): 331–350.
Yoo, F., Jung, H.J. and Oh, K.W. (2021), 'Motivators and barriers for buying intention of upcycled fashion products in China', *Sustainability*, 13(5): 2584.
Zhao, X., Korey, M., Li, K., Copenhaver, K., Tekinalp, H., Celik, S., Kalaitzidou, K., Ruan, R., Ragauskas, A.J. and Ozcan, S. (2022), 'Plastic waste upcycling toward a circular economy', *Chemical Engineering Journal*, 428: 131928.
Zhou, Y., Zhang, Y. and Li, X. (2021), 'Upcycling of paper waste for high-performance lithium-sulfur batteries', *Materials Today Energy*, 19: 100591.

Chapter 12

Designing with Waste in Hong Kong

Kevin Cheung and Kyungeun Sung

Introduction

Kevin Cheung is a product designer in Hong Kong who has been practising upcycling for the past decade. He and his design team have created all kinds of electronic products, including speakers and flashlights. While working, he realized that what the industry required was different from what he learned at the university. What the university taught him was all about critical thinking, research and evidence-based design, creative problem solving, out-of-the-box ideas, new product development from scratch and design for social and environmental good. Every project in the industry, however, seemed to come and go in a very similar way. Designers received components or parts (or specifications) and designed pretty boxes outside, which was the standard procedure for new product development. He recalled the initial launch of his product and went to the store to locate it. In the electronics section of a Hong Kong-based retailer, he saw a whole wall of speakers of different sizes and colours. When he looked at it – the wall with his tiny contribution to it – he asked himself, 'Why do we still need to design speakers?'. The answer: because the company's objective was to generate profits. Designers are tasked with creating new and slightly improved products that will sell well. The issue arises when consumers upgrade to new speakers, their old ones are typically sent to landfill in Hong Kong. There are only three landfills in Hong Kong, and they are consistently being filled up, as reported by the news media every year. Every day, 18,000 tonnes of waste goes into landfill. We need to find somewhere else for landfills; in fact, we are building incinerators to burn the rest of the waste. This is unsustainable as a lot of what people throw away takes hundreds or even thousands of years to decompose. Moreover, there is a lot of valuable material in there. Kevin wondered how much of it could be used to make new products. This led him to his upcycling journey.

As part of his current work, Kevin and his team have been picking up exhibition carpets from a nearby convention centre. New exhibitions open every week. When they set up, the new carpets are cut to fit each booth, and there are lots of small

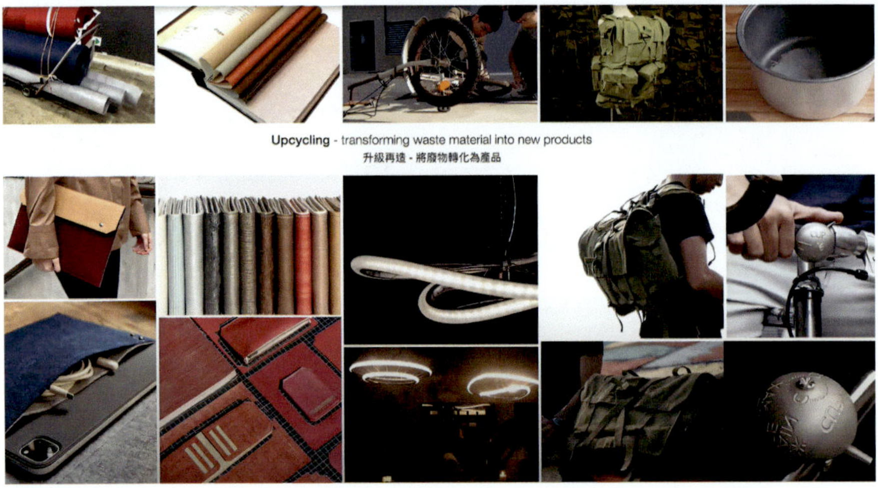

Figure 12.1 New products through upcycling by Kevin Cheung.

pieces of carpet left over in different colours. These small pieces are too small to reuse, so Kevin's team collects them. They sew carpet pieces together to make cases for gadgets (see Figure 13.1, first column). They also collect old or outdated wallpaper sample books or booklets and turn them into wallets (see Figure 13.1, second column). Another material they collect are broken bicycle wheels (those that are rusted and bent that cannot be normally reused or repaired). Kevin's team used these unusable wheels, replaced the tyres with LEDs and made them into ceiling lights (see Figure 13.1, third column). The team also used old military bags, changed the structure, and reorganized the straps to make them more usable (see Figure 13.1, fourth column). The team also used the inner pots of the rice cooker, which are no longer used for health and safety reasons due to the significant scratches. They found that the resonance of the inner pots was good, so they pressed the pots into tiny bells for bicycles (Figure 13.1, last column).

Boombottle is the best example of Kevin's upcycling work. Therefore, the following section is dedicated to describing the Boombottle project in detail.

Boombottle

Boombottle was Kevin's first upcycling project (or upcycled product). Based on his previous job designing speakers, he knew that any container could be turned into a speaker box to make it resonate. Kevin collected medicine bottles from various pharmacies and clinics, used the internal volume of plastic bottles to make them resonate, turned the bottle cap into a volume control, and put a light inside (see Figure 13.2).

Kevin wanted to design an entire production chain to handle a large volume of this waste material for mass production. He therefore worked with a plastic recycler

Figure 12.2 Boombottle design by Kevin Cheung.

to collect all sorts of plastic bottles. For the fabric, he used the small carpet pieces from new exhibitions at the local convention centre. He made the Boombottle speakers in a sheltered workshop that employed skilled disabled people. Kevin simplified the creation process and taught people with disabilities how to make a Boombottle speaker by hand. The entire process, from material collection to production, has been carried out locally in Hong Kong (see Figure 13.3).

Each Boombottle uses around 500g of material that would otherwise go to landfill. The direct environmental impact may not be significant (compared to Hong Kong's 18,000 tonnes of daily waste). However, the portable speaker's distinctive design and strong sound quality have generated positive word-of-mouth. Many customers contacted Kevin, inspired by Boombottle's story. The product's main goal is to promote upcycling and share a narrative that is worth repeating. The story has gone viral, and various media outlets have featured Boombottle and other upcycled products by Kevin over the past decade (see Figure 13.4). Boombottle has also recently been included in the permanent collection of the M+ Museum (one of Hong Kong's most prestigious design museums).

Workshops, Installation Arts and Exhibitions

Since Boombottle's success, Kevin has seen Hong Kong's citizens, schools, and businesses change their perceptions, beliefs and opinions about waste and upcycling; they have also become more open to participating in upcycling projects. Kevin's team, therefore, developed and delivered more public engagement activities, events, and exhibitions to further influence the people of Hong Kong. The first educational workshop Kevin's team delivered was for adults to make xylophones

Figure 12.3 Boombottle local production process in Hong Kong.

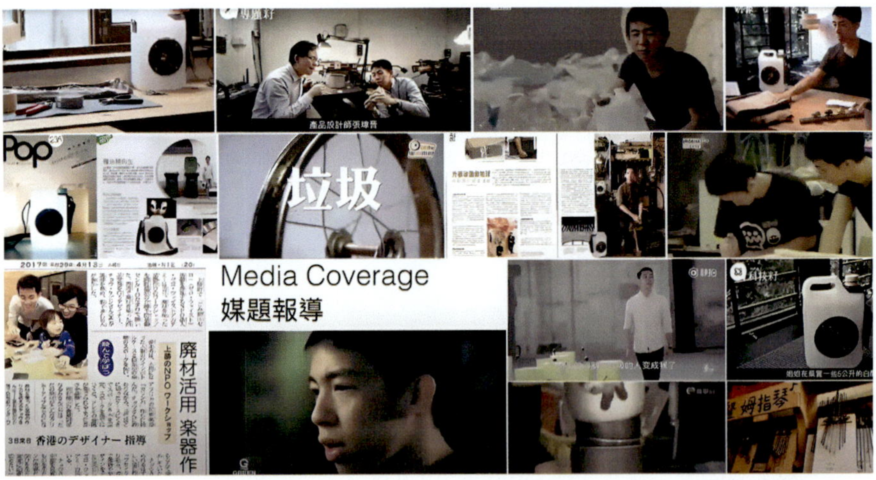

Figure 12.4 Media coverage on Boombottle and other upcycled products by Kevin Cheung.

from waste plastic Coke bottles (collected from Coca-Cola's factory in Hong Kong). They then organized a workshop for children. As Hong Kong can be very hot in the summer, everyone carries a small fan. The children's educational workshop used these small portable fans and waste plastic bags to create small inflatable monsters (see Figure 13.5 top). The next adult upcycling workshop was to make kalimbas using old umbrella ribs (see Figure 13.5 bottom).

In addition to educational workshops, Kevin's team has worked with various companies, NGOs, and governments to upcycle their waste into something meaningful. For example, they had one street installation where the team asked community members to collect, clean and bring Tetra Paks (plasticized carton packaging) after they had drunk their milk, juice, etc. They taught them how to press the Tetra Paks into small windmills and these were then hung along the streets of Hong Kong. They started by making a few hundred small windmills that people could see hanging on the streets; however, people brought more Tetra Paks for further involvement (Figure 13.6).

Another street installation project was with Greenpeace who held an exhibition along the Harbourfront in Hong Kong to raise awareness of single-use plastic bags. Kevin's team worked on the temporary shelter installation made from hundreds of plastic laundry bags. Most homes in Hong Kong are very small, and many people do not have a washing machine. People in Hong Kong, therefore, often take their clothes to launderettes and, for reasons of hygiene, use a new plastic bag to take the clean clothes home after each wash. After the first use, the perfectly reusable plastic laundry bags are thrown away. These bags are durable and have the same translucent

Figure 12.5 Inflatable monsters (top); and Kalimba making process (bottom).

Figure 12.6 Tetra Pak windmill street installation.

white colour with different branding graphics. Kevin's team used irons to melt and join them together to create a giant piece and build a temporary shelter (5m x 5m x 2m) (Figure 13.7 top). Another public engagement project was with the government who wanted to create a sustainable, upcycled bicycle that would allow cyclists to interact with picnickers in a meaningful and entertaining way. Kevin's team created a live music platform by mounting an old piano on a tricycle. The old piano was sixty years old and had been made in Hong Kong. It belonged to a family with two children who had grown up and no longer wanted to play the piano. Kevin's team picked it up, refurbished it and worked with a local craftsman to create the moving piano platform (Figure 13.7 bottom).

The team has worked on a project with a French company which sells luxury goods such as leather bags and silk scarves decorated with beautiful and delicate

Figure 12.7 Laundry bag temporary shelter (top); and moving live music platform (bottom).

Figure 12.8 Silk scrap tree making and installation.

patterns and colours. Kevin's team went to the company's factory in Paris and collected all the scraps of silk that were too small to be used in any of its products. As the silk scraps were very wrinkled, they used a bamboo embroidery ring to stretch them out so that people could see their beautiful patterns. Kevin worked with a team of local women to make 5,000 of these rings and then constructed a tree out of them. The upcycled silk scrap tree was a pop-up installation in the company's main Hong Kong store (Figure 13.8).

Discussion and Conclusion

This chapter is based on Kevin's presentation in the Seminar on Upcycling in Asia in October 2022 as part of the International Upcycling Research Network funded by UKRI (UK Research and Innovation) AHRC (Arts and Humanities Research Council). The work described in this chapter is not exhaustive. Kevin has a much more extensive portfolio of other exciting, and fascinating projects in and around upcycling, and the list is growing by the day. If you visit his website (http://kevin-cheung.com/), you can see all the work Kevin has done over the last decade.

Overall, Kevin's work is an inspiration to all the product and upcycling designers and makers across industries and countries. More importantly, Boombottle and other upcycled products solve one of the critical problems in commercializing upcycled products – scalability (Pal et al., 2021; Singh et al., 2019). Most upcycled products are one-offs based on craft upcycling (e.g., Holdsworth, 2024). They are not necessarily suitable for mass production for the mass market, as they start from very different materials and are based on manual work without machines for standardization (Singh et al., 2019). By involving people with disabilities in the production process, Boombottle is not only environmentally sustainable, but also has clear benefits for local people and society, in line with the growing trend of using upcycling as a strategy to benefit both the environment and disadvantaged groups (Descubes et al., 2018; Pusz et al., 2024).

Many upcycling designers and makers tend to move away from making commercial products for the mass market and become artists or craftspeople for commissions or luxury one-offs, either by choice or due to a lack of infrastructure, equipment and skills (e.g., Sung and Cooper, 2015). To have a significant positive impact on the environment, economy and society, more scalable upcycling ideas and concepts should be sought and developed in the future, with a systems thinking/approach (or product supply chain consideration) involving material suppliers and production site (just like Kevin's Boombottle). Everyone has their waste to throw away or upcycle. In this sense, Kevin believes that there are no limits to upcycling. It is for everyone and it can be done everywhere. We hope that this chapter will inform and inspire students, teachers, and professionals in the creative industries around the world to join us on our journey to spread the word about upcycling and make a real difference.

Acknowledgements

This work was funded by the AHRC Research Networking grant for the International Upcycling Research Network [grant number AH/W007134/1]. The writing of this book chapter was possible with the generous Research Award from DMU. We thank Nicky Harding, Stuart Lawson, Kate Cheyne, Kelley Wilder, Siobhan Keenan, Christine White, Jill Cowley, Deborah Cartmell, Mike Kagioglou, and Katie Normington for supporting this research at DMU.

References

Descubes, I., McNamara, T., & Cragg, T. (2018), 'Recycling waste and upcycling people: A new type of environmentally-motivated social enterprise strategy', International Journal of Manufacturing Technology and Management, 32(3): 270–296.

Holdsworth, C. (2024), 'Bedding into bags: the life histories of materials, makers and the time of making in a case study of fabric upcycling', Social & Cultural Geography, 25(3): 423–440.

Pal, R., Samie, Y., & Chizaryfard, A. (2021), 'Demystifying process-level scalability challenges in fashion remanufacturing: An interdependence perspective', Journal of Cleaner Production, 286, 125498.

Pusz, M., Jonas, A. E., & Deutz, P. (2024), 'Knitting circular ties: Empowering networks for the social enterprise-led local development of an integrative circular economy', Circular Economy and Sustainability, 4(1): 201–232.

Singh, J., Sung, K., Cooper, T., West, K. and Mont, O. (2019), 'Challenges and opportunities for scaling up upcycling businesses – The case of textile and wood upcycling businesses in the UK', *Resources, Conservation and Recycling*, 150: 104439.

Sung, K. and Cooper, T. (2015), 'Sarah Turner – Eco-artist and designer through craft-based upcycling', *Craft Research*, 6(1): 113–122.

Chapter 13

Organic Waste Upcycling for Materials Design: A New Framework Proposal Based on Food Waste

Luca Alessandrini, Valentina Rognoli and Karen Marie Hasling

Introduction

In contemporary economies, there is a growing recognition of the significance of circular economy practices, which involve upcycling waste and turning by-products into valuable materials and products (Campbell-Johnston et al., 2020). This shift towards upcycling represents a pivotal element in transitioning from a linear to a circular and potentially more sustainable model of production (Braungart & McDonough, 2013). At the heart of this paradigm are materials and their design, as well as their development through the upcycling of waste and by-products, which play a crucial role throughout the product lifecycle. The design field emphasizes the role of materials in the creative process (Rognoli et al., 2010). Approaches like DIY-Materials (Rognoli et al., 2015) and methodologies such as Material Driven Design (Karana et al., 2015) provide designers with essential tools to work with materials, enabling them to create specific material experiences (Hekkert & Karana, 2014). The design phase has been found to determine up to 80 per cent of a product's environmental impact (Thackara, 2005). Therefore, the adoption of a materials-focused design methodology places a greater responsibility on designers to embrace a sustainability perspective and adopt a design for sustainability approach (Ceschin & Gaziulusoy, 2016). Interestingly, it is only in recent years that designers have started to delve into deeper, more profound impacts and life cycles of materials (Pollini et al., 2021), indicating a renewed interest among the new generation of designers to develop circular solutions (Rognoli et al., 2021). As a result, many designers around the world are actively working on materials with the aim of creating more sustainable alternatives (ibid.), often using upcycling to minimize the use of new raw materials and reduce the impact at the end of their life (Bofylatos, 2022).

Materials encompass the entire chain, from the extraction of raw materials to the multiplicity of manufacturing processes (Vezzoli, 2014) and serve as the foundation for product creation. As such, designers working with materials bear significant responsibility, as their choices can substantially influence the overall environmental impact of products. As many designers continue to evolve their practices, the upcycling of organic waste (OW) to create materials, in particular, can provide a further benefit, as it has a lower environmental impact in the disposal phase due to its biodegradability (Lasaridi et al., 2011): up to nine times lower environmental impact for most of the ISO14040:2006 impact categories in the disposal phase compared to non-biodegradable waste (David et al., 2020). In this regard, with organic waste, we intend biodegradable waste to have an organic origin. They can be in solid, semi-solid (with a high percentage of liquid) or fully liquid form, all of which have in common the ability to be broken down naturally by living organisms (EU Commission, 2020). A previous study (Rognoli et.al., 2023) has investigated how designers are at the forefront of developing materials that upcycle one of the most significant streams of OW: Food Loss and Waste (FLW). This occurs throughout the whole Food Supply Chain (FSC) since, globally, approximately a third of all food produced for human consumption is lost or wasted: the Food and Agriculture Organization of the United Nations (FAO, 2019) estimates that approximately 14 per cent of all food produced is lost at the post-harvest stage.

By understanding the similarities between practice-based activities on upcycling OW, other designers and companies can be inspired to adopt sustainable practices and create similar outputs by upcycling different waste streams. The research aims to envision a practical approach to promoting the potential scalability of materials as a design practice fostering positive impact trends.

Framework Development

In an attempt to provide a valuable tool for inspiring and analysing new practices, we created a framework that aims to reframe case studies from the perspective of practitioners and designers. Although there are many ways to design frameworks that can contain, organize and classify qualitative data from case studies (Cwalina et al., 2020), this study used Rognoli et al.'s (2023) recent framework, which could serve as a valuable tool for designers and companies interested in developing circular materials that upcycle OW including FLW. The previously proposed framework (Rognoli et al., 2023) formed the basis of the extended version presented here (Figure 14.1), which has the same aim: to provide both a ready-to-use visual database tool for representing case-study profiles and for collecting data on them, allowing for a more comprehensive and in-depth analysis.

Section 1 collects the case study's general information:

- *Food supply chain phase*: indicates from which phase of the FSC the FLW originates.
- *Year*: represents the year of the project's realization.
- *Designer/brand*: name of the designer or the upcycling brand that developed the case study.

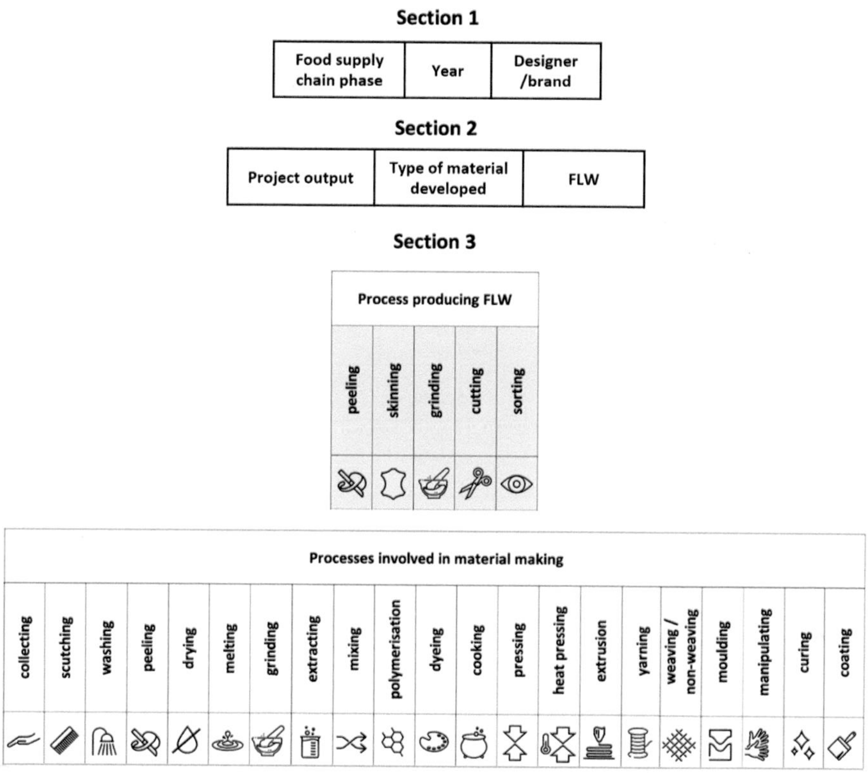

Figure 13.1 Extended FLW framework to analyse case studies.

Section 2 shows all the information regarding the downstream (FLW used), midstream (materials developed) and upstream (project outputs) of every case study (Wahl et al., 2008).

- *Project output*: specifies whether the designer or the upcycling brand released a new product or developed a new material using FLW.
- *Type of material developed*: demonstrates how the FLW was transformed into fillers, bioplastics, pigments, fabrics, leathers, non-woven textiles, etc.
- *FLW*: wastes coming from the FSC from the designers/brands.

Section 3 includes the processes indicated with easily recognizable icons regarding two aspects:

- *Process producing FLW*: summarizes five generic processes corresponding to the FSC phases, including peeling, skinning, grinding, cutting, and sorting.
- *Process involved in material making*: includes a collection of processes involved in making the materials used for the project outputs by the designers. These

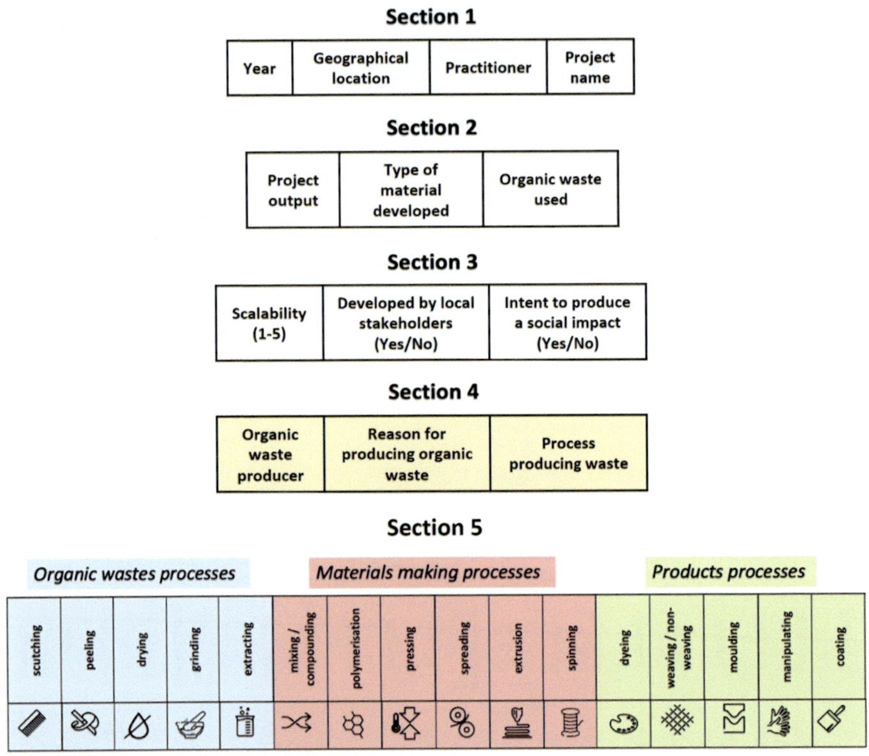

Figure 13.2 New framework proposal.

processes include scutching, washing, drying, melting, mixing, polymerization, dyeing, pressing, extruding, weaving/non-weaving, moulding and more.

We expanded the framework further by adopting some new classification criteria so that we can include any OWs, not limited to FLW (see Figure 14.2).

Section 1 collects general case study information:

- *Year*: year of the case studies development.
- *Geographical location*: where the case study was developed.
- *Practitioner*: name of the practitioner who developed the case study.
- *Project name*: name given to the project by the practitioner.

Section 2 is about the information regarding the aspects of upcycling:

- *Project output*: same as the earlier description.
- *Type of materials developed*: same as the earlier description.
- *Organic waste used*: every OW coming from multiple waste streams beyond FSC.

Section 3 was not present in the previous FLW framework. It focuses on three aspects that have been considered by the practitioners themselves in their projects, according to the findings of the previous study (Rognoli et.al, 2023):

- *Scalability*: In the previous study (Rognoli et.al, 2023), it was noted that the level of scalability between the case studies was very different. So here we propose to measure its level from 1 to 5, which could help to understand the level of potential impact that the cases will be able to create with their practices. Here, impact refers to the ability to use larger quantities of OW and, consequently, the ability to produce and share higher quantities of materials and products adopting it (Bofylatos, 2022). This goes along with the possibility of upcycling and the impact created through that. Level 1 indicates the case at the stage of proof-of-concept (materials/products realized but no intent for further replication). Level 2 indicates a limited series of production (e.g., a few copies of a product or modest quantities). Level 3 indicates reproducible production and intent for further production but through crafting only. Level 4 indicates replication through the combination of crafting and limited mass production techniques. Level 5 indicates replication at the industrial scale.
- *Developed by local stakeholders (Yes/No)*: By collecting this information, we would understand if the practitioners were developing their projects with the intent to source OW or part of the processes involving stakeholders. This was a factor stated by the practitioners among the collected case studies in the previous FLW framework. It is to understand the importance of localism among the principles applied by practitioners. The nature of the involved local stakeholders can vary since localism can imply two potential effects: (i) sustainability-wise: reduction of emissions for transportation; and (ii) social-wise: potentially bringing economic and/or social benefits to local communities (Santagata et al., 2021).
- *Intent to produce a social impact (Yes/No)*: This information indicates the practitioners' willingness to create a positive social impact. The FLW framework has multiple ways of doing this (e.g., involving local farmers or populations in the manufacturing activities, recovering disappearing traditional crafts, using waste that causes problems to communities). This was considered an important aspect to collect as it shows how different aspects of sustainability can be combined through upcycling.

Section 4 collects information on different aspects of OW production. Similarly to the FLW framework presented previously, it starts with the collection of processes producing OW. At the same time, in this extended version, more typologies of information can be collected on who the OW producer is and the reason for the production. This aims to contribute to building a profile of the OW producer.

- *Organic waste producer*: Collecting information on the entities producing OW contributes to picturing a more detailed scenario of the case studies. This could offer the chance to trace profiles of who produced OW and to benefit in finding possible redundancies along the collection, possibly spotting valuable insights. Based on the knowledge collected in the previous work (Rognoli et al., 2023), these entities are by-products of other companies' processes and waste from cultivation and domestic kitchens. However, this expanded framework aims to go beyond FLW.

- *Reason for producing organic waste*: OW can be produced for many reasons, including industrial processes, expired goods, overproduction, unwanted goods, etc. Tracking this aspect would lead to the possible consideration of recursive leaks in OW production while spotting OW that could be reused.
- *Process producing waste*: OWs are the result of processes coming from different sectors, locations, streams and stakeholders, which are managed by practitioners with the intention of giving these wastes other purposes (e.g., household waste, agricultural harvesting, fishing industry processing). Differently from the FSW framework, a few changes were made. It was noted that if this new proposed framework aims to include case studies working with as many OWs as possible, the waste-generating processes could not have been included only in the categories of the previous FLW framework, which only refers to the five given generic actions (peeling, skinning, grinding, cutting, and sorting). For this reason, a column has been left to give a brief description of each process that produces OW.

Section 5, like the previous FLW framework, is entirely dedicated to the processes that practitioners have undertaken to achieve the outputs of their projects, as well as the processes that produce OWs. Some of the 20 processes from the previous FLW framework have been excluded, as they are similar actions, resulting in 16 processes in the new framework. One process has been added: spreading. This process is widely applied to create flat materials using polymers such as alternative leathers. The selected processes were divided into three:

- *Organic waste processes*: processes to transform the OW to make it usable to realize materials, including scutching, peeling, drying, grinding and extracting.
- *Material-making processes*: processes to realize materials, including mixing/compounding, polymerization, pressing, spreading, extrusion and yarning.
- *Product processes*: processes to implement the realized materials into products, including dyeing, weaving/non-weaving, moulding, manipulating and coating.

Application of the Framework: Selected Case Studies

To concretely demonstrate how the proposed framework can be used to analyse and communicate practices of organic waste upcycling, this section presents a curated selection of case studies. Each case was selected to illustrate a specific combination of organic waste source, material development, production process and intended impact. The projects exemplify how design practices can promote

Year	Geographical location	Practitioner	Project name	Project output	Type of material developed	Organic waste used	Scalability (1-5)	Developed by local stakeholders Yes/No	Intent to produce a social impact Yes/No	Process producing waste	scutching	peeling	drying	grinding	extracting	mixing / compounding	polymerisation	pressing	spreading	extrusion	spinning	dyeing	weaving / non-weaving	moulding	manipulating	coating
2014	Italy	Orange Fiber	Orange Fiber	orange peel made fabric	yarn	orange peel	4	yes	no	industrial food processing					x	x					x					
2019	Italy	High Society	Lamps	lamps from beer leftovers	filler	beer leftovers	3	no	no	industrial food processing			x	x		x								x		
2014	Philippines	Carmen Hijosa	Pinatex	pineapple leaves leather	alternative leather	pineapple leaves	5	yes	yes	agricultural harvesting	x		x			x			x				x			x
2019	UK	Lucy Huges	MarinaTex	plastic from fish skin	biopolymer	fish skin	3	yes	no	industrial food processing					x		x									

Figure 13.3 Case studies examples in the framework.

circularity not only through material innovation but also by engaging local stakeholders and addressing social challenges.

By mapping these real-world practices through the lens of the extended framework, we aim to highlight key parameters such as scalability, the role of local actors, and the intentional generation of social value. Together, these case studies offer a multidimensional view of upcycling as a design strategy – one that connects material experimentation, sustainability, and community-driven innovation. They also serve as inspiration for designers, researchers and industries seeking to replicate or evolve similar approaches.

Case Study 1: Piñatex

Developed by Spanish designer Carmen Hijosa, *Piñatex* is one of the most iconic examples of upcycling agricultural waste in the fashion sector. Originating in the Philippines, the project reuses pineapple leaves – typically discarded during harvest – to create an alternative leather. Through scutching, drying and felting processes, the fibres are transformed into a non-woven material that mimics the texture of traditional leather.

The production of Piñatex is not only environmentally conscious but also socially inclusive: it involves local farmers' cooperatives in the Philippines, supporting rural economies and enhancing community-based production. The material has reached high levels of scalability (Level 5), being adopted by global brands such as Adidas and Stella McCartney.

- *Practitioner*: Carmen Hijosa
- *Location*: Philippines / Spain
- *Year*: 2015
- *Organic waste used*: pineapple leaves (agricultural waste)
- *Material developed*: alternative leather
- *Project output*: Used in fashion products (Adidas, Stella McCartney)
- *Scalability*: 5

- *Local stakeholders involved*: Yes
- *Social impact intent*: Yes.

Case Study 2: Orange Fiber

Founded in Italy in 2016, *Orange Fiber* is a pioneering startup that transforms orange juice industry by-products into sustainable textiles. Using cellulose extracted from orange peels – wasted during industrial juicing processes – the company creates a lightweight fabric that can be blended with silk.

The process combines traditional spinning with bio-based innovation, aligning industrial efficiency with material experimentation. Collaborating with fashion giants like H&M and Ferragamo, Orange Fiber has achieved industrial scalability while fostering local partnerships and aiming to generate positive social impact. The project sets a strong benchmark for closed-loop design in the textile industry.

- *Practitioner*: Orange Fiber
- *Location*: Italy
- *Year*: 2016
- *Organic waste used*: Orange peels from juice industry (food processing)
- *Material developed*: Silk-blended sustainable fabric
- *Project output*: Fashion collections (H&M, Ferragamo)
- *Scalability*: 5
- *Local stakeholders involved*: Yes
- *Social impact intent*: Yes

Case Study 3: MarinaTex

MarinaTex is a bioplastic developed in the UK by designer Lucy Hughes as a response to marine waste generated by the seafood industry. Using fish scales and skins – by-products of local fish processing – she created a translucent, flexible material ideal for packaging. Its composition, based on natural binders such as agar, allows it to biodegrade within six weeks in a home compost environment.

This project demonstrates the power of low-tech, bio-based innovation in replacing synthetic materials. Though still in a pre-industrial stage (Level 4 scalability), MarinaTex has gained international recognition, winning the 2019 James Dyson Award. The project also fosters local stakeholder involvement and aims to reduce marine pollution while stimulating eco-entrepreneurship.

- *Practitioner*: Lucy Hughes
- *Location*: United Kingdom
- *Year*: 2019
- *Organic waste used*: Fish scales from seafood industry
- *Material developed*: Bioplastic for packaging
- *Project output*: Fashion accessories
- *Scalability*: 4

- *Local stakeholders involved*: Yes
- *Social impact intent*: Yes

Case Study 4: Spent Coffee – High Society

High Society, an Italian design studio based in the Dolomites, has developed a series of lighting products using spent coffee grounds collected from local cafés. This organic waste, typically discarded after beverage preparation, is rich in lignin and cellulose, making it suitable as a filler in composite materials.

The studio transforms the coffee grounds through drying, grinding and mixing processes, combining them with bio-resins to form mouldable materials used in lighting design. The resulting lamps showcase a distinctive texture and scent, embodying the poetic potential of waste while contributing to a circular material aesthetic.

The project shows a limited reproducibility (Level 3 scalability), relying on local sourcing and artisanal production methods. While not explicitly targeting social impact, High Society's approach highlights the role of small-scale, regionally grounded initiatives in valorizing urban organic waste through design.

- *Practitioner*: High Society
- *Location*: Italy
- *Year*: 2019
- *Organic waste used*: Spent coffee grounds
- *Material developed*: Lamps
- *Project output*: Consumer products and design objects
- *Scalability*: 3
- *Local stakeholders involved*: Yes
- *Social impact intent*: No

Discussion and Conclusion

The new extended version of the framework presented in this chapter aims to obtain more insightful results than those previously developed. The extended framework's new section 3 (Figure 14.2) measures three correlated factors: scalability, localism and social impact willingness. This section was inspired by the previous study (Rognoli et al., 2023), indicating that practitioners considered these factors across case studies. From this perspective, using this extended framework could lead to projects that focus on social impact aspects as a rationale, becoming case studies of socially sustainable upcycling. Consequently, creativity is likely to be seen not only in the realization of sustainable materials but also in the inventiveness to make them socially sustainable. In many cases, localism is linked to sustainability through local resources or processes that reduce transport emissions (Santagata et al., 2021) and is linked to social impact (Chrispim et al., 2023). Section 4 gathers information on different aspects of OW production to contribute to building a profile of OW producers. It starts by including the processes that produce OW, going beyond the food loss and waste (FLW) framework (Rognoli et al., 2023) to include other sectors and streams. It is possible to ascertain an expanded profile of OW producers through process details, producer entities and production reasons which, in turn, allows for the identification of redundant OW production and usage opportunities. This leads to the formation of clusters of unutilized OW, which in turn suggests the reasons for its creation may be linked with sectors, OW typologies and so forth, with the potential for further upcycling possibilities. Section 5 of the new framework is dedicated to the processes that the practitioners perform to obtain their project outputs (materials or products). This section proposes an expanded version of section 3 of the FLW framework (Rognoli et al., 2023), presenting none of the five processes producing FLW (relegated to a new dedicated column) and showing a reduced number of processes (16 instead of 20) describing the practitioners' actions. Moreover, these processes have been divided into three different phases (OW processing, materials making and product making). These updates were implemented following an analysis of the processes conducted in the previously conducted FLW case studies collection. Some processes were redundantly executed to create the same typology of materials: for example, to create compounds having OW fillers inside, the OW needs to be processed, ground, and then, to realize the final material, they must be mixed/compounded with a polymer or a binder. The organic fibres are usually scutched and spun to realize yarn out of OW fibres. On the other hand, the processes for obtaining products made from materials from OW vary from one product to another. Realizing a lamp, for example, can be done in a wide variety of ways. In the end, indicating such generic processes will facilitate the classification of a diverse array of case studies. This revised section aims to provide a straightforward, illustrative representation of the steps required to replicate the case study.

The proposed framework provides a starting point and analytical framework for collecting and analysing cases of upcycling organic waste that could inform further research and applications across sectors. Documenting diverse projects spanning waste sources, processes, materials, and products enables the identification of patterns, challenges, and scalability. This collection could benefit researchers working on sustainability issues such as the circular economy, waste valorization, materials engineering and industrial ecology. The framework could also provide practical knowledge for industrial stakeholders involved in waste generation, handling or valorization. Producers facing disposal costs could refer to analogous cases to assess the integration of new upcycling pathways. Waste processors and producers could use access to documented applications to assess market viability. Documentation of barriers, enablers and social impacts could provide a context for policymakers seeking to develop regulations and transition programmes. The framework could also mobilize the public and non-profit sectors to support community initiatives based on the links between upcycling, equity and economic resilience. Finally, the processes, materials, and products that are realized could inspire creative entrepreneurs, designers, and organizations to explore local waste valorization. By highlighting ongoing innovation, this framework aims to stimulate further bottom-up experimentation through replicable methodologies and the identification of under-explored spaces. The new framework is a ready-to-use framework for other practitioners who are willing to upcycle OW that does not just come from the FSC. It offers insights into the development of products from different OW and the processes involved in transforming waste into materials and products. Its novelty lies in its adaptability to different types of OW while collecting and classifying information to visually communicate potential applications. The framework provides an infrastructure for high-impact research on OW reuse. By building a comprehensive case-study collection using the framework, the authors will be able to obtain qualitative data on trends in successful material innovation across different waste sources and projects. Tracking scalability, localism and social impact can highlight the conditions needed to take waste-based solutions from concept to industry. Documenting sources and processes could enable the replication of successful practices for under-utilized OW. The framework is intended to help researchers identify high-potential waste applications and foster collaboration to promote circularity among waste producers.

References

Bak-Andersen, M. (2021), *Reintroducing materials for sustainable design: Design process and educational practice*. Abingdon: Routledge.

Bofylatos, S. (2022), 'Upcycling systems design, developing a methodology through design', *Sustainability*, 14(2): 600.

Campbell-Johnston, K., Vermeulen, W.J., Reike, D. and Brullot, S. (2020), 'The circular economy and cascading: towards a framework', *Resources, Conservation & Recycling*, X(7): 100038.

Ceschin, F. and Gaziulusoy, I. (2016), 'Evolution of design for sustainability: From product design to design for system innovations and transitions', *Design studies*, 47: 118–163.

Chrispim, M.C., Mattsson, M. and Ulvenblad, P. (2022), 'The underrepresented key elements of Circular Economy: A critical review of assessment tools and a guide for action', *Sustainable Production and Consumption*, 35: 539–558.

Cwalina, K., Barton, J. and Abrams, B. (2020), *Framework design guidelines: conventions, idioms, and patterns for reusable. net libraries*. Boston: Addison-Wesley Professional.

David, G., Croxatto Vega, G., Sohn, J., Nilsson, A.E., Hélias, A., Gontard, N. and Angellier-Coussy, H. (2021), 'Using life cycle assessment to quantify the environmental benefit of upcycling vine shoots as fillers in biocomposite packaging materials', *The International Journal of Life Cycle Assessment*, 26: 738–752.

Desing, H., Braun, G. and Hischier, R. (2021), 'Resource pressure – a circular design method', *Resources, Conservation and Recycling*, 164: 105179.

European Commission (2020), 'Biodegradable waste'. Available at: https://environment.ec.europa.eu/topics/waste-and-recycling/biodegradable-waste_en (accessed 14 June 2023).

Food and Agriculture Organization of the United Nations (2019), 'The state of food and agriculture. Moving forward on food loss and waste reduction', Available at: https://ec.europa.eu/food/system/files/2019-10/fw_lib_fao-2019_en.pdf (accessed 15 June 2023).

Grossoehme, D.H. (2014), 'Overview of qualitative research', *Journal of Health Care Chaplaincy*, 20(3): 109–122.
Hekkert, P. and Karana, E. (2014), 'Designing material experience', in E. Karana, O. Pedgley, V. Rognoli (eds), *Materials Experience: Fundamentals of Materials and Design*, 3–13, Oxford: Butterworth-Heinemann.
Hijosa, C.A.A. (2015), *Piñatex, the Design Development of a New Sustainable Material.* London: Royal College of Art.
Jensen, P.D., Basson, L., Hellawell, E.E., Bailey, M.R. and Leach, M. (2011), 'Quantifying 'geographic proximity': Experiences from the United Kingdom's National Industrial Symbiosis Programme', *Resources, Conservation and Recycling*, 55(7): 703–712.
Karana, E., Barati, B., Rognoli, V. and Zeeuw Van Der Laan, A. (2015), 'Material driven design (MDD): A method to design for material experiences', *International Journal of Design*, 9(2): 35–54.
Lasaridi, K. and Stentiford, E. (2011), 'Upcycling'organic waste in a world of thinly distributed resources', *Waste Management & Research*, 29(11): 1115–1116.
McDonough, W. and Braungart, M. (2010), *Cradle to cradle: Remaking the way we make things*. Berkeley: North Point Press.
Pandit, P., Annaldewar, B.N., Nautiyal, A., Maiti, S. and Singha, K. (2020), 'Sustainability in fashion and textile. Recycling from Waste in Fashion and Textiles: A Sustainable and Circular Economic Approach', in P. Pandit, S. Ahmed, K. Singha, S. Shrivastava (eds), *Recycling from Waste in Fashion and Textiles: A Sustainable and Circular Economic Approach*, 177–198, Austin: Scrivener Publishing.
Parisi, S., Rognoli, V. and Sonneveld, M. (2017), 'Material Tinkering. An inspirational approach for experiential learning and envisioning in product design education', *The Design Journal*, 20(sup1): S1167–S1184.
Pedgley, O., Rognoli, V. and Karana, E. (2021), *Materials experience 2: Expanding territories of materials and design*. Oxford: Butterworth-Heinemann.
Pollini, B. and Rognoli, V. (2021), 'Early-stage material selection based on life cycle approach: tools, obstacles and opportunities for design', *Sustainable Production and Consumption*, 28: 1130–1139.
Rognoli, V. and Parisi, S. (2021), 'Material Tinkering and Creativity', in L. Clèries, V. Rognoli, S. Solanki, P. Llorach (eds), *Materials Designers. Boosting talent towards circular economies*, 20–25, Barcelona: Elisava.
Rognoli, V., Alessandrini, L. and Pollini, B. (2023), 'From food waste to circular materials for design: Experimenting with matter from unconventional origins', in R. Earley and R. Hornbuckle (eds), *Design Materials and Making for Social Change*, 8–29, New York: Routledge.
Rognoli, V., Garcia, C.A. and Pollini, B. (2020), 'DIY Recipes: Ingredients, processes & materials qualities', in L. Clèries, V. Rognoli, S. Solanki, P. Llorach (eds), *Material Designers: Boosting talent towards circular economies*, 27–33, Barcelona: Elisava.
Rognoli, V., Bianchini, M., Maffei, S. and Karana, E. (2015), 'DIY materials', *Materials & Design*, 86: 692–702.
Rognoli, V., Petreca, B., Pollini, B. and Saito, C. (2022), 'Materials biography as a tool for designers' exploration of bio-based and bio-fabricated materials for the sustainable fashion industry', *Sustainability: Science, Practice and Policy*, 18(1): 749–772.
Santagata, R., Ripa, M., Genovese, A. and Ulgiati, S. (2021), 'Food waste recovery pathways: Challenges and opportunities for an emerging bio-based circular economy. A systematic review and an assessment', *Journal of Cleaner Production*, 286: 125490.

Thackara, J. (2005), *In the bubble. Designing in a complex world*, Cambridge MA: MIT Press.
Vea, E.B., Romeo, D. and Thomsen, M. (2018), 'Biowaste valorisation in a future circular bioeconomy', *Procedia Cirp*, 69: 591–596.
Vezzoli, C. (2014), 'The "material" side of design for sustainability', in in E. Karana, O. Pedgley, V. Rognoli (eds), *Materials Experience: Fundamentals of Materials and Design*, 105–121, Oxford: Butterworth-Heinemann.
Wahl, D.C. and Baxter, S. (2008), 'The designer's role in facilitating sustainable solutions', *Design issues*, 24(2): 72–83.

Part IV

People and Community

Chapter 14

UPCYCLING THROW-AWAY PEOPLE: LESSONS FROM UPCYCLING RESEARCH

Joon Sang Baek and Kyungeun Sung

Introduction

When we discuss upcycling in an academic setting, by upcycling, we usually mean material processes that transform waste or used products, components and materials into a new product or material of a higher quality or value than the constituent elements, often in the context of the circular economy. As an alternative to traditional mass production and consumption based on virgin materials (within the current linear economy of take, make, use and dispose of), upcycling contributes to waste reduction, improved material/resource efficiency, reduced industrial energy consumption and therefore reduced carbon emissions, the creation of new job opportunities and people's psychological/emotional wellbeing (Singh et al., 2019; Sung, 2017; Sung et al., 2019; 2021). Indeed, it is important to address our waste, resource, pollution and emissions issues in the present moment, as reflected in several UN Sustainable Development Goals (e.g., UN SDG 11, 12 and 13). Provided that we endeavour for upcycling, circular economy and sustainable development ultimately for people's benefit, it would also be beneficial to consider how upcycling could be used to address other aspects of human life beyond the material production and consumption domains. Based on the concept of social upcycling (Wegener & Aakjær, 2016), we aim to continue a discussion on expanding the scope of objects of upcycling to immaterial wastes, especially underutilized human capabilities including knowledge, creativity, empathy, care, goodwill and a sense of community.

Social Upcycling

According to Wegener and Aakjær (2016) social upcycling is in line with the ongoing expansion of sustainability issues, including not only environmental but also social matters. This trend makes the term upcycling relevant when engaging with potentially marginalized communities, including prisoners, immigrants, refugees, the elderly and unemployed youth. These individuals are frequently defined by deficiency and the societal challenges they may present (ibid.). Defined as 'the act of recognising and nurturing potential [human] value and insisting on dignity and creative expression as guidelines for human interaction' (ibid., 3), it produces greater values out of the resources such as political recognition and support, voluntary labour and philanthropic commitment. The broad notion of sustainability inherent in social upcycling is related to 'an increasing need to embrace the complexity and entangled socio-materiality of innovation' (ibid., 13). The authors argue that to achieve a greater sustainability impact; there is a growing need to embrace the complexity of entangled social-ecological dimensions of innovation in design for sustainability, including upcycling design. This argument underpins our proposition that innovation through upcycling design needs to incorporate a new type of resource for innovation that grows most abundantly on the planet – social resource.

In this chapter, we raise the following question: What, if anything, can we learn from upcycling design knowledge to design for empowering undervalued human resources such as throw-away people? The context of the study is the ageing population in South Korea, one of the most rapidly ageing societies among G20 members and likely the most aged society among the advanced economies in 2050 (OECD, 2019). Using a case study, we question if the capabilities of an ageing population are undervalued resources that are subject to upcycling design and explore the possibility of learning from the established knowledge of upcycling design in developing a designerly approach to empowering elders.

Generation of Throw-Away People in Rapidly Ageing Society

With the rapid ageing of South Korea's population, an increasing number of individuals are leaving the workforce or have stopped actively looking for work due to age-related factors, despite possessing both the willingness and necessary qualifications to contribute to the labour market. According to Cahn (2004), the author of *No More Throw-Away People*, they are 'throw-away people'. As one of the most rapidly ageing societies among the Organisation for Economic Co-operation and Development (OECD) countries (OECD, 2016), South Korea is facing a surging senior population that is prone to poverty, social isolation, loneliness and depression (Yi & Hwang, 2015). This means a huge loss to both individuals and society due to the reduced tax revenue, increasing social welfare costs, wasted human resources, and other social opportunity costs of unemployment (Johansson & Kriström, 2022).

In societies with a high level of old-age income poverty, retired people are economically vulnerable and exposed to poverty and related social problems. According to an OECD report, South Korea's elderly poverty rate is the highest among OECD members by a large margin. In terms of the old-age income poverty rate, which is measured by the percentage of people aged over 65 having income lower than 50 per cent of the median equivalized household disposable income, South Korea's elderly poverty rate was 43.4 per cent, followed by Latvia (39%), Estonia (37.6%), and Mexico (26.6%) (OECD, 2021). Public efforts by the government to address poverty among seniors have not been effective (Lee, 2021). These problems can be mitigated by empowering seniors eager to re-engage socially and economically. Furthermore, uplifting undervalued seniors can yield numerous social, economic, and cultural advantages. These include fostering a sense of community and interconnectedness, enhancing the mental, emotional, and physical well-being of the elderly, facilitating cross-generational learning in workplaces, alleviating the strain on social services, preserving the wisdom and experience of older generations, promoting age diversity and inclusivity and presenting a positive view of ageing.

One example is Time Banking, a non-monetary market system that operates based on community self-help. Edgar Cahn started Time Banking in the 1980s with the aim of filling the gaps in public social services when the Reagan administration was pushing cuts to spending on social programmes. He believed that we are all assets and have something valuable to share with others. This belief became one of the core values of Time Banking, a global initiative of exchanging reciprocal service that uses time as currency and is active in 26 countries (Cahn, 2004). Another example is the Community Senior Club in South Korea, a civilian organization which aims to provide occupational training and creating job opportunities for people aged over 65. The increasing number of senior people in Korea are left alone because young people tend to move out of their parent's house after they get married. Therefore, aged parents need to be more independent to make a living and maintain their quality of life. However, finding a job after retirement is not easy as they feel burdened by the workload and are not welcomed by employers (Kim et al., 2015). In this context, the Community Senior Club was established by the Korea Labour Force Development Institute for the Aged an affiliated institute of the Ministry of Health and Welfare in 2004. The subject of our case is a local branch of the club located in Ulsan, South Korea's seventh-largest metropolitan city, where industrial complexes and large and small farms coexist.

Case Study: Ulsan Senior Club

Ulsan Senior Club supports the senior people in the region in participating in social activities by providing occupational training and creating job opportunities. There are two types of job programmes available for senior people in the club: social service and market-orientated service. In the former, senior citizens participate in social and public activities such as public facility management, traffic guidance in the child safety zone and caring for the elderly who live alone. They offer services that create societal values but not necessarily profits; in return, they are paid by the senior club. The salary is approximately $300 USD per month. Because the programme is run solely on the government budget, the number of available positions depends on the budget amount; this changes on a yearly basis and is also limited. The contract is renewed every nine months.

On the contrary, the market-orientated service programme, as the name implies, aims at offering marketable services by utilising the knowledge, skills, and competencies of elderly people. Under the programme, a dozen business models operate in the form of for-profit organisations. Elderly people operate them with support from the centre, and any profits are reinvested into the business. Compared to the social service programme, the market-orientated service programme is economically more sustainable because it creates more jobs, offers its workers more stable working conditions once the business is on track and enhances feelings of attachment and self-esteem. For the senior club, creating and supporting as many marketable business models as possible is the highest priority.

Supply/Value Chain Approach to Interventions for Scaling Up Upcycling

Upcycling design methods and processes depend on the context and object of design; as Wegener and Aakjær (2016) mentioned, there is no one-size-fits-all methodology. We chose the value-chain approach to upcycling design inspired by systems thinking because we found it appropriate to be applied to complex sociotechnical problems. This value-chain approach contributes to diagnosing the challenges of scaling up upcycling with a systemic perspective that expands the scope of diagnosis to the entire value chain. It views upcycling as a sociotechnical process that involves multiple stakeholders along the product lifespan. Hence, we applied the value-chain approach to implement and expand the services offered by the Senior Club.

The supply or value-chain ecosystem of upcycling businesses has varied stakeholders, including material suppliers, upcycling designers and makers, retailers, and consumers. These different stakeholders have distinctive challenges for scaling up upcycling. For example, according to a UK study focusing on the fashion and furniture sectors (Singh et al., 2019), challenges for material suppliers in sourcing materials include the lack of materials, complex legislation, lack of funding, and risks and financial implications. The challenges for upcycling designers and makers include a lack of affordable resources (space, time, equipment, skills and materials), extensive marketing of upcycled products, and a lack of materials with suitable quality and quantity. The challenges for retailers in selling upcycled products include a lack of resources for the suitable marketing of upcycled products, outreach of upcycled products, consumers' negative perception of upcycled products and financial constraints. The challenges for consumers in buying upcycled products include the availability of good-quality products, affordable products and the accessibility of upcycled products. To address these challenges, a number of interventions (or success factors) were identified (see Table 16.1) (Singh et al., 2019).

Based on the case study, we identified the following challenges encountered by the senior club in implementing and expanding their programmes:

Table 14.1 Success factors for scaling up upcycling (Singh et al., 2019)

Material supplier	Designer/maker	Retailer	Consumer
Raising awareness for material donation, provision of legislative and financial support	Provision of critical resources (e.g. skilled staff) and financial support for skills development, work environment and marketing	Effective marketing, positive consumer awareness, and improved quality of products	Improved awareness of upcycled products, better purchasing experience (e.g. one-stop shop), and lower price of upcycled products

- *More supply than demand*: An increasing number of seniors are looking for jobs, but there are not as many opportunities available. Compared to previous generations, baby boomers tend to be more independent from their children both socially and economically. Thus, they have a greater need for financial independence.
- *Limited budget*: There are limited funds allocated to the social services programme which has limited social impact. The senior club pays the social service workers and the number of job openings depends on the budget allocated to the programme. To generate more jobs and a social impact, the club needs to expand and scale-up its market-orientated services.
- *Developing viable market-orientated service models*: These services create jobs for the elders and profits for the senior club. The organisation wants to create more job opportunities for seniors by creating new success cases and scaling up existing ones. However, the organization's ability to develop and grow competitive service models is limited because its entire staff consists of social workers.
- *Matching seniors' capabilities and business opportunities*: One of the more successful cases of the market-orientated service is the Spirit, which makes and delivers Korean traditional foods for ancestral ceremonies. It has successfully turned the elders' life-long experience in performing ancestral rites in their families into a solid business model. It has been running for more than a decade and has positive customer responses in terms of the quality of the food and service (Kim et al., 2015). The club needs to develop more success stories by understanding seniors' capabilities and market demands.
- *Scaling up success cases*: Scaling up an innovation requires a different set of knowledge and skills different from creating one: an understanding of supply and demand, drawing out and articulating the core of the innovation, and procuring the right skills and resources to execute a growth strategy (Gabriel, 2014), which the senior club or the seniors may not possess.

The challenges that the senior club encountered cannot be solved internally and require collaboration with external partners along the empowerment process (or value chain). In this regard, we found the value-chain approach useful for

Table 14.2 Success factors for facilitating the Senior Club's services

Senior club	Designer/business consultant	Service provider (seniors)	Customer
Collaboration opportunities with external partners in the areas of demand (e.g., new service development), raising awareness of the club activities in local communities	Provision of opportunities to collaborate with the senior club, provision of knowledge, skills and financial incentives for new service/business model development	Effective marketing, positive customer awareness, and improved quality of products/services	Improved awareness of the services offered by seniors, a shift from competing on price to competing on quality and customer experience

developing strategies to implement and expand the services offered by the senior club. Stakeholders involved (or need to be involved) in this process include the senior club, service designer/business consultant, service provider (seniors), and customers. We brainstormed interventions aimed at addressing the problems identified in the previous section (see Table 16.2).

The demand for the empowerment of underutilized or 'throw-away' people is growing in ageing societies (Dugarova, 2017). Thus, a systematic approach to empowering them and helping them transform into social constituents of greater value is needed. In this chapter, we argued that the upcycling concept can be extended to the realm of human resources and used to empower people who are no longer economically active despite their willingness and capabilities. Building on the concept of social upcycling, we explored the potential of applying upcycling design to the empowerment of seniors devalued in contemporary society. The case study suggests that upcycling design and empowerment design share several similarities and differences.

In terms of their goal, both approaches seek to add value, whether it is underused products or undervalued individuals. They rely on the creativity of designers in the transformation process. However, specific knowledge necessary for this value-adding process may differ depending on the context of design. For example, the methodologies of both approaches are adaptive, based on context and available resources. Upcycling design tends to be product centric, while empowerment design has a broader scope encompassing services, systems, strategies and governance. Therefore, typical design approaches to the former come from product and fashion designs and the latter from participatory, service, system and strategic designs. Upcycling Design and Empowerment Design, while both seeking to add value, primarily differ in their focus areas – environmental vs societal. The impact of each design approach caters to distinct areas of improvement (see Table 16.3).

This study was driven by the hypothesis that upcycling design knowledge can inform and benefit design for empowering undervalued human resources. It led us to the conclusion that this is true and so can vice versa. That is, the notion of upcycling human resources can expand the scope of upcycling design beyond the

Table 14.3 Comparison of upcycling design and empowerment design

	Upcycling design	Empowerment design
Goal	Adding quality or value to waste, used or underused products	Developing the power of marginalised people
Transformation	Transforming the tangible and material qualities of something to assign it greater value using designers' creativity	Transforming people and/or their environment to empower them socially, politically or economically using designers' creativity
Design object	Product, material, built environment, etc.	Product, service, system, process, strategy, governance, etc.
Related concept	Circular economy, (environmental) sustainability	Social innovation, (social) sustainability
Design approach	Product design, fashion design	Participatory design, service design,
Methodology	It depends on the contextual needs, skills and resources; nature of design objects; no dedicated methodology	It depends on the contextual needs, skills and resources; the design approach used; no dedicated methodology
Motivation	Out of need and for the environment	Out of need and for society
Impact	Waste reduction, improved material/resource efficiency, reduced industrial energy consumption, reduced carbon emissions, the creation of new job opportunities and people's psychological/emotional well-being (Singh et al., 2019; Sung, 2017; Sung et al., 2019; 2021)	Improvement of life conditions, citizen engagement and democratic participation in political and social matters, social innovation and problem-solving, enhanced quality of life, and strengthening of the user's position (Ertner et al., 2010)

materiality of innovation and ecological sustainability. Sustainability issues we face are increasingly complex and require a holistic approach to addressing multi-faceted problems that are neither social, material, nor ecological, but all of them combined. Limitations in achieving sustainable development with insular solutions (e.g., products) mean the limited impact of upcycling design. On the other hand, human resources are abundant, and the problems and solutions generated by people are limitless. This suggests more opportunities for upcycling design.

Some critical academics, particularly those in the physical sciences and engineering, may argue that the definition of upcycling and what it encompasses is already too broad. The common criticism is that upcycling covers too many concepts and practices, such as innovative or advanced forms of repair, reuse, refurbishment, upgrading, redesign, remanufacturing, recycling, etc., as does another umbrella term, the circular economy. Therefore, this new perspective of including immaterial resources as objects of upcycling is subject to further criticism. However, new terms are constantly being coined and introduced; it is up to the users to decide how to use or not use the term (i.e., depending on the social practice of collective language use). Only time will tell if upcycling is a useful concept to describe our attempts to utilize immaterial waste for higher value(s) that contribute both to society and the economy. For now, we think, why not? We invite other academics to join the discussion.

Conclusion

This chapter explored the potential of applying upcycling design knowledge to empowering undervalued elders based on a case study. It remains to be explored what other knowledge than the value-chain approach can be relevant to the expanded concept of upcycling. Besides, the scope and scale of the case study are limited. Our hypothesis thus needs to be tested in a broader context. We acknowledge that the problematic situation framed in this study is not universal. In many countries with advanced social welfare systems, such as the UK, seniors typically do not need to nor do they want to work in the later part of their life. The scope of undervalued or throw-away people, however, varies in context and can be found anywhere: unemployed graduates, retired professionals, stay-at-home parents, people with disabilities, skilled immigrants, ex-convicts, and those whose skills are not fully utilized in their current jobs. The capabilities of this cohort are as abundant as their diversity and the impact of design will be limitless.

References

Cahn, E. S. (2004), *No More Throw-Away People: The Co-Production Imperative*, Washington DC: Essential Books.

Dugarova, E. (2017), 'Ageing, Older Persons and the 2030 Agenda for Sustainable Development', *United Nations Development Programme*, Available at: https://www.undp.org/publications/ageing-older-persons-and-2030-agenda-sustainable-development

Ertner, M., Kragelund, A. M. and Malmborg, L. (2010), 'Five enunciations of empowerment in participatory design', *Proceedings of the 11th Biennial Participatory Design Conference*, 191–194.

Gabriel, M. (2014), 'Making it big. Strategies for scaling social innovations', *nesta*, Available at: https://www.nesta.org.uk/report/making-it-big-strategies-for-scaling-social-innovations/

Johansson, P.O. and Kriström, B. (2022), 'On the social opportunity cost of unemployment', *Journal of Economic Policy Reform*, 25(3): 229–239.

Kim, S., Pahk, Y. and Baek, J. S. (2015), 'Senior Club in the district of Jung-gu, Ulsan', in Y. Lee, P. Moore (eds), *Ageing and Ingenuity*, Hong Kong: DESIS Network.

Lee, J.H. (2021), 'S. Korea ranks worst in OECD for poverty in old age,' *Hankyoreh*, Available at: https://english.hani.co.kr/arti/english_edition/e_national/1013419.

OECD (2021), 'Pensions at a Glance 2021', *OECD iLibrary*, Available at: https://www.oecd-ilibrary.org/finance-and-investment/pensions-at-a-glance-2021_ca401ebd-en.

Singh, J., Sung, K., Cooper, T., West, K. and Mont, O. (2019), 'Challenges and opportunities for scaling up upcycling businesses–The case of textile and wood upcycling businesses in the UK', *Resources, Conservation and Recycling*, 150: 104439.

Sung, K. (2017), *Sustainable production and consumption by upcycling: Understanding and scaling up niche environmentally significant behaviour*, Nottingham: Nottingham Trent University.
Sung, K., Cooper, T. and Kettley, S. (2019), 'Developing interventions for scaling up UK upcycling', *Energies*, 12(14): 2778.
Sung, K., Singh, J. and Bridgens, B. (2021), *State-of-the-art upcycling research and practice: Proceedings of the international upcycling symposium 2020*, Cham: Springer.
Wegener, C. and Aakjær, M. (2016), 'Upcycling – a new perspective on waste in social innovation', *Journal of Comparative Social Work*, 11(2): 242–260.
Yi, E.S. and Hwang, H.J. (2015), 'A study on the social behavior and social isolation of the elderly Korea', *Journal of Exercise Rehabilitation*, 11(2): 125–132.

Chapter 15

Exploring Successful Cases of Community Participation in Waste Management and Upcycling in Thailand

Montida Lertnimanoradee, Busayawan Lam and Eujin Pei

Introduction

The amount of solid waste generated in Thailand in 2022 was 25.7 million tonnes (70,411 tonnes) per day. This is considered a serious issue since the amount of municipal waste in Thailand was the second highest compared to those produced by other ASEAN (Association of Southeast Asian Nations) countries (Environment, 2017). The amount of solid waste has increased by 3% compared with that in 2020 (Pollution Control Department, Pollution Control, 2022). Several factors, such as the lack of cooperation in waste sorting before disposal and limited knowledge of solid waste management within households, caused the mishandling of solid waste at the community level (Maschamadol, 2018). Due to increasing waste problems, the Thai government has acknowledged the importance of this issue and set the solid waste solution as a national agenda (Buranasingh, 2017). Although Thailand has various laws and regulations in relation to the waste-management process, which describe specific duties and responsibilities of local government organizations in regard to the collection, storing and transportation of waste (including exploitation of waste, such as electric power business and investment promotion), the lack of integration and participation of various sectors, especially at the earlier stages of the process (e.g., waste sorting), makes them less effective. Other strategies (e.g., encouraging people to reduce the amount of waste from their consumption and promoting environmentally-friendly packaging) have not been officially established – these include providing clearer knowledge on waste management to the general public (Mokhsart, 2022). At present, the 3Rs principle (Reduce, Reuse, Recycle) has been used as an operating framework for waste management, which can be divided into three phases: upstream (reduce), midstream (reuse), and destination (recycle). The aim is to reduce the amount of waste, promote waste segregation at source, create an efficient waste collection system, and properly dispose (Ministry of Interior, 2022).

The People's Constitution of 1997 stated in section 76 that: '*The state must encourage and support the participation of people in policy and political decision making, planning for economic, social and political development, including monitoring the use of state power at all levels*.' The constitution emphasized the importance of citizen participation, especially in the decentralization of power from the central state to local organizations elected by the people, including local communities. People in the community can participate in projects to solve their community problems and act in the majority of their daily lives economically, socially and culturally, which allows them to think about and make decisions regarding these issues. It is crucial that local people create a database of community information and analyse community needs to evaluate the value within the community context by themselves. Therefore, the self-reliant community plan preparation process was offered to community members, which aims to enable people in a community to have a chance to meet, talk, express their opinions and brainstorm ideas regarding problems and challenges facing their communities (Na Pomphet, 2007).

The idea of getting citizens involved in the problem-solving process is not new to Thailand. People in small communities have already been given the opportunity to participate in the consultation process on how to deal with serious matters, such as those related to drugs, economics, environment and infrastructure, welfare and social issues. In 2004, the *Handbook of Community Based Solid Waste Management* (CBM) was used to get people involved in the waste-management process at community level. This includes (i) recognizing waste problems in their areas; (ii) finding solutions; and (iii) deciding on waste-management methods that are suitable for their community. In this way, people can take action according to an agreed plan created by them. Waste management projects are still carried out at community level, such as waste banks, waste markets, waste donations, and composting from organic waste projects. (Pollution Control Department, Community Based Solid Waste Management, 2019).

This chapter aims to discuss the challenges and issues related to MSW (Municipal Solid Waste) management in Thailand, highlight the potential of upcycling as a solution, emphasize the importance of citizen participation in waste management, propose ways to improve waste-management practices at community level and emphasize the need for more effective and sustainable waste-management strategies. By incorporating upcycling and encouraging citizen involvement, the chapter seeks to promote a more holistic approach to waste management that empowers local communities and fosters responsible waste-handling practices.

Methods

This chapter presents a preliminary research study conducted as part of a doctoral research project, focusing on community participation, participatory design, and upcycling practices in waste management. The study followed a systematic approach, beginning with an extensive literature review to explore existing knowledge on these topics within Thai small communities.

The literature review involved analysing scholarly articles, government reports, municipal reports and subdistrict official websites, primarily sourced from Thai Journals Online (ThaiJO) and official community-based resources. Key search terms included 'Community participation in waste management', 'Community-driven upcycling in Thailand', and 'Challenges and opportunities in upcycling in Thai communities'. The literature review aimed to identify effective participatory strategies, community engagement models, and upcycling practices relevant to small Thai communities. The key findings highlighted existing strategies and best practices that Thai communities implement in waste management and upcycling, providing a foundation for further investigation.

To complement the literature review, the research incorporated exploratory case studies, utilizing secondary data from online sources, including municipal reports, government databases and community-driven initiatives. The case-study approach followed Yin's (2017) framework for case-study research, allowing for an in-depth examination of how participatory design and upcycling are applied in different community settings. Table 16.1 summarizes the three key research areas

Table 15.1 Summary of key research areas and supporting literature

Research Area	Key Findings	Key References
Community participation in waste management	Community-led waste management varies based on governance structures and available resources. Effective models integrate education, incentives, and local policies. Leadership engagement is crucial for long-term sustainability. Participation frameworks include structured models (decision-making, operations, evaluation and benefit-sharing).	Chainam & Bouphan (2022), Pollution Control Department (2019, 2022), Ministry of Interior (2022), DEQP (2023), Maschamadol (2018), Atthaporn et al. (2024)
Community-driven upcycling in Thailand	Co-design and participatory workshops are essential in community-led upcycling. Traditional handcrafting and cultural preservation play a major role in upcycling acceptance. Knowledge-sharing initiatives enhance upcycling adoption and local innovation. Economic incentives support the sustainability of upcycling projects.	Chiangka (2018), Nimthong et al. (2024), Yin (2017), Warin Chamrab Municipality (2018), Maschamadol (2018), Thanangngam Subdistrict Administration (2024), Phetthirasawat (2023)
Challenges and opportunities in upcycling in Thai communities	Lack of funding and limited training restrict upcycling scalability. Cultural heritage and locally sourced materials present untapped opportunities. Motivation and tangible benefits (economic and environmental) drive participation. Market demand and policy support influence long-term sustainability.	Limsomboon et al. (2024), DEQP (2023), UN Environment Programme (2017), Phetthirasawat (2023), Thungluknok Subdistrict Administration (2024)

derived from the literature review, aligning with the case study findings. This categorization was developed based on existing frameworks in participatory design, waste management and community-based upcycling, ensuring a systematic approach to data analysis.

Results

The following sub-sections describe four case studies. Each case study begins with a summary of the key activities followed by key lessons learned. While three case studies involve direct upcycling practices, the second case study (Yommarat Village) is included to provide insights into community participation models, waste-separation strategies and behavioural engagement – critical foundations for successful upcycling initiatives. By understanding the challenges in fostering participation, particularly in the absence of upcycling, this case study highlights barriers that need to be addressed to implement community-driven circular economy models effectively.

Case Study 1: Tunglooknuk Sub District, Nakhon Pathom Province

In the Tunglooknuk sub-district, the 3Rs principle has been effectively implemented for waste management. Firstly, the community successfully raised awareness about waste reduction. Secondly, they improved recycling practices by increasing the number of bins and promoting waste sorting campaigns. This addressed improper waste disposal which is a significant contributor to household waste issues. Lastly, the community embraced upcycling design, creating valuable products from scrap materials, leading to additional income sources and job opportunities for residents. The combined efforts to reduce, reuse and recycle waste have made the waste-management approach in this community both sustainable and economically beneficial. Upcycling practices also support recycling activities as they encourage all waste to be sorted properly (Chiangka, 2018). The community participation activities of the Tunglooknuk sub-district can be grouped into four steps: (i) participation in decision-making, (ii) participation in operation, (iii) participation in receiving benefits, and (iv) participation in evaluation (Parepuno, 2018).

The first key lesson (Lesson 1) learned was *clear communication regarding the roles and responsibilities of people in the community in each step of participation.* The Thungluknok sub-district administrative organization divided community participation activities into four steps to make them easier for people to participate in and provided clear guidelines for each step. The organization invited members of the community to analyse, search and define problems on waste issues themselves. After that, related agencies (both in the public and private sectors) provided further information to help people make decisions, which led to the selection of suitable activities and the elimination of inappropriate ones. Community members were encouraged to share knowledge regarding waste management. In this way, they would be able to comment on waste-management targets/objectives and provide recommendations for appropriate operations. The advantages of participatory activities were clearly communicated to community members (Limsomboon, Likitpanyarat, & Srisawang, 2024). For example, carrying out activities together gave people a chance to get to know each other and learn how to work collaboratively. These participatory activities enabled people to take positive action to address the root cause of the problems (improper disposal of household waste) and improve their quality of life. Other members of the community also acknowledged their contributions. Furthermore, these participatory activities encouraged community members to express their opinions and give suggestions on how to improve the operations regarding waste management in the government sector (Figure 16.1). The methods of encouraging people in the community to participate included organizing training/seminars and enabling community leaders to take charge of solid waste-management activities. By participating in decision-making, operation and evaluation, all members received benefits from these positive actions together (Thungluknok Subdistrict Administration, 2024).

The second key lesson (Lesson 2) was that *the activities aimed at creating upcycled products should align with the cultural and lifestyle aspects of individuals to effectively motivate their participation in waste management.* Recognizing the significance of community members' culture, lifestyle, values, local customs, and attitudes is crucial when designing activities that cater to their needs and preferences. The small rural community has a strong familiarity with traditional handcrafting techniques, which they utilize to create various everyday items such as Thai bath towels, clothing, and home decor (Phetthirasawat, 2023). These handcrafted products hold significant cultural and practical value for the community, representing a tangible connection to their heritage and way of life. The transmission of handcraft techniques across generations has played a significant role in preserving cultural roots within each locality, encompassing narratives, processes, patterns, and heritage techniques. This approach acknowledges the significance of considering the socio-cultural context in which individuals operate. By incorporating elements that reflect their values, aesthetics, and preferences, they create a sense of familiarity and relevance. The Thungluknok community, for example, initiated an upcycling activity involving the manual weaving of UHT beverage boxes into baskets, which served as functional containers for individuals to carry items when visiting temples or markets. In order to foster engagement and

Figure 15.1 The community agencies consistently organize educational activities to promote household waste separation within the community (Tunglooknuk, 2016).

active involvement in waste-management practices, it is crucial to ensure that the activities related to the creation of upcycled products resonate with the cultural values and lifestyle choices of the target audience. By tailoring these activities to align with people's cultural backgrounds, traditions and daily routines, we can enhance their motivation to participate in waste-management initiatives.

Case study 2: Yommarat Village, Tha Nang Ngam Subdistrict, Phitsanulok Province

The 3R principle guided the waste-management activities implemented in the Yommarat village. The primary objective of these activities was to address the specific waste-related concerns and requirements of the local population. The key issue identified within the community was the inadequate disposal of waste, characterized by practices such as waste burning and improper disposal of garbage outside designated areas, leading to detrimental environmental pollution. The study focused on helping people in the community understand the guidelines so that they could dispose of household waste more efficiently. The local organization started by providing basic knowledge about waste sorting to encourage people to separate waste and manage each type of waste properly. In addition, the local organization followed the community participatory model proposed by the workshop to solve community problems by listening to people's opinions and suggestions regarding waste management (Figure 16.2). The community participatory model on waste management at Ban Yommarat can be grouped into three steps: (i) participation in the planning process, (ii) participation in the operations of waste-management activities, and (iii) participation in the evaluation process. According to the evaluation of the planning process, six key considerations were identified, namely: budget saving, ease of implementation, availability of materials in the community, possibility, participation of people in the community, and benefits for residents (Thanangngam Subdistrict Administration Organization, 2024).

The third key lesson (Lesson 3) was from *the local organization focused on the promotion of waste-separation practices by providing explicit guidelines for each specific type of waste management.* The sorting of waste serves as the initial and crucial step in community-led waste-management initiatives. Differentiating waste types is facilitated by the implementation of a four-colour system: green bins

Figure 15.2 Tha Nang Ngam Subdistrict Administrative Organization's meeting to address waste management, seeking participation from the local community (Tha Nang Ngam, 2023).

for compostable waste, yellow bins for recyclable waste, blue bins for general municipal waste, and red bins for hazardous waste (DEQP, 2023). To assist community members in comprehending the various waste categories, the local organization in Yommarat Village facilitated a range of activities. For instance, they conducted workshops on transforming compostable waste into fertilizers and biofertilizers, which not only helped villagers reduce the expense of purchasing compost but also contributed to the generation of income for families by encouraging the separation and sale of recyclable waste to recycling centres (Figure 16.3) (Nimthong, Sinchao, Patthamasonti, Sirinakhon, & Faiyue, 2024).

The fourth lesson (Lesson 4) was that *the presentation of the benefits arising from community activities in waste management is imperative to foster active participation and engagement among individuals.* Although six key considerations were identified at the planning stage, the local organization did not integrate them into waste-management activities. They struggled to involve people in participatory activities, particularly regarding benefits for local residents. While Yommarat Village did not incorporate upcycling, its community engagement strategies and waste-separation efforts are valuable for understanding the broader context of participatory waste management. The challenges in participation observed here, particularly the lack of perceived benefits, highlight a key issue that can also hinder upcycling adoption in similar communities. Addressing these barriers through clear communication, incentives and structured participation models could improve future upcycling initiatives by ensuring communities are aware of the direct benefits. This case study provides a critical contrast to the more successful upcycling-driven participation seen in other cases, reinforcing the importance of engagement at all stages of sustainable waste-management projects. Since the 'participation in receiving benefits' was not clearly communicated, this case ultimately demonstrated only three stages of community participation. Motivation plays a pivotal role in influencing individuals to take action, and external motivation is closely tied to the anticipation of rewards and reciprocal benefits (Assegaff, Kurniabudi, & Fernando, 2016). Therefore, greater emphasis on the

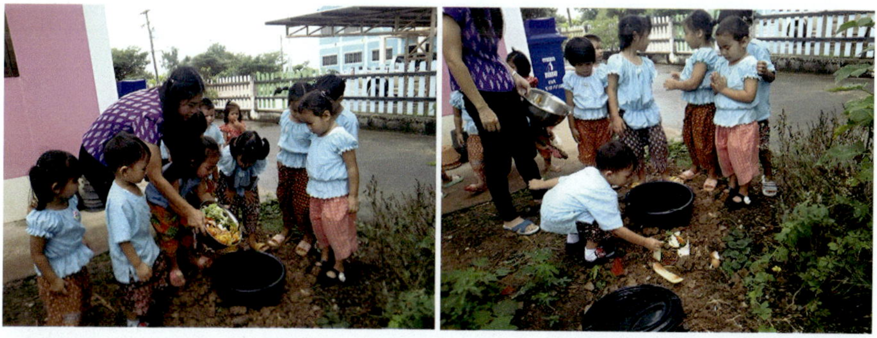

Figure 15.3 Tha Nang Ngam Subdistrict Administrative Organization, in collaboration with the Children's Centre, organized an activity to make fertilizer from wet waste (Tha Nang Ngam, 2023).

'participation in receiving benefits' could have enhanced awareness of the advantages of community-led waste management. In this case, benefits could have included reducing household costs by producing compost, generating income through the sale of recyclable waste and improving quality of life by reducing local pollution and waste accumulation.

Case Study 3: Waeng Yai Sub District, Khon Kaen Province

The local organization in the Waeng Yai sub-district adopted the 3Rs as the guiding principle for waste management, aligning with the community participation guidelines provided by the Thai government (Atthaporn, Kittikronwaranon, Tarasi, & Yolai, 2024). Central to their approach was the cultivation of a sense of community belonging and active engagement in operational responsibilities throughout the waste-management process, encompassing six stages: (i) participation in defining the problem, (ii) planning and decision-making, (iii) operation; (iv) receiving benefits; (v) evaluation; and (vi) responsibility. Since the local organization concentrated on managing and reducing waste, five additional key areas were set as main targets: (i) reducing the amount of solid waste generation; (ii) solid waste sorting; (iii) solid waste collection; (iv) solid waste collection and transportation; and (v) solid waste disposal (Chainam & Bouphan, 2022).

The fifth lesson learned (Lesson 5) is *the importance of utilizing design knowledge within waste-management workshops to develop a diverse range of innovative waste-management approaches.* Despite the clear definition of the five targets, the specific plans pertaining to the application of the

3Rs remain unclear. However, it was observed that the integration of design knowledge and skills could significantly enhance waste-management practices, particularly in the context of reuse. This involves the transformation of discarded materials into valuable products, which presents opportunities for community members to participate and receive associated benefits actively. At the Waeng Yai sub-district, creative approaches to waste management were employed, such as the hand weaving of bags and hats using coffee packaging by an elder group, the utilization of vinyl banners to create aprons, the repurposing of refill bags from detergent and fabric softeners into bags, and the conversion of discarded rice cookers into biomass stoves (Figure 16.4) (Warin Chamrab, 2018).

Figure 15.4 Waengyai villagers participate in training on waste processing design and knowledge exchange with the Warin Chamrab Group (Warin Chamrab, 2018).

Case Study 4: Wiang Phrao Sub District, Chiang Mai Province

Chiang Mai province followed the National Master Plan regarding waste management by applying the 3Rs principle and the Community Solid Waste-Management Action Plan. Both plans required cooperation from communities and local governments to move towards a zero-waste community. In line with these plans, the local organization incorporated research findings from academic and researcher-led studies on community participation in waste management. By leveraging these insights, they adapted their waste-management processes and developed activities tailored to suit the specific needs and characteristics of their local communities. The solid waste-management process, guided by community participation, encompasses seven strategies aimed at encouraging active engagement within the activities. These strategies include:

(a) raising awareness among community members to foster their participation in waste-management efforts;
(b) developing and enhancing the capacity of community leaders to effectively address solid waste challenges;
(c) addressing waste issues by categorizing specific target groups within the community;
(d) community participation process involving community members and local agencies in the area;
(e) implementing a system for monitoring, corrective measures, and inspections;
(f) promoting the roles and responsibilities of local agencies in waste management; and
(g) creating incentives and inspiration to stimulate behavioural changes among community members.

These activities are connected to the process of community participation, which can be divided into four steps: (i) the planning and decision-making stage, (ii) operational activities, (iii) receiving the benefits, and (iv) the evaluation process (Wiang Phrao Municipality, 2023).

The sixth lesson (Lesson 6) was that *preparing and enhancing the capacity of design skills is an integral part of the activities related to participation in the operational aspects of waste management.* The local organization implemented activities aimed at developing and enhancing the capacity of community leaders, including village headmen, headman's assistants, municipal council members, village committees, volunteers, health volunteers, the heads of the elderly and housewives. The activities involved conducting training workshops to address solid waste problems. The workshops focused on educating participants about upcycling and adding value to products through the cultivation of creative-thinking skills. Interested individuals and organizations were invited to participate in these workshops, which aimed to promote waste transformation. The creative activities specifically targeted the participation of community leaders, members, and local agencies within the area. This led to the establishment of a sub-district action group responsible for organizing various projects and activities in the community. These initiatives facilitated the implementation of waste-management practices, such as the conversion of organic waste into fertilizer or bio-fertilizer, the repurposing of scrap materials from packaging into home-decor items, and the processing of agricultural waste into rice straw papers.

The seventh lesson (Lesson 7) was that *the presentation of the benefits derived from participating in waste-management activities assumes a significant role in motivating individuals engaged in activities.* The local organization in Wiang Phrao incentivizes good behaviour in waste sorting and acknowledges outstanding waste-management practices at the sub-district level through the provision of awards. This initiative is aimed at motivating individuals to enhance their awareness of waste-sorting practices. Additionally, by highlighting the benefits of effective waste management and showcasing innovative approaches to waste reduction, inspiration is generated among individuals to adopt creative waste management strategies. There is a strong emphasis on presenting the benefits resulting from participation in waste-management activities, as this serves as a crucial motivator for individuals involved in these activities. By highlighting the positive outcomes, individuals are more motivated and actively participate in waste-management efforts. These benefits encompass various aspects, including environmental improvements, pollution reduction, enhanced public health, economic opportunities through waste recycling and upcycling, as well as an overall enhancement in the quality of life for community members.

Discussion and Conclusion

These four case studies demonstrate how community participation models and the 3Rs principle (Reduce, Reuse, Recycle) were adapted to the distinct contexts of small communities in Thailand. Each case study highlights tailored waste-management practices accompanied by community participation frameworks aimed at fostering collaboration between individuals and local organizations to promote sustainable waste management. Across the case studies, community participation was found to follow three to six sequential stages depending on the local context, including: (i) problem identification and planning, (ii) operational activities, (iii) evaluation, and (iv) participation in receiving benefits. These participation models provided valuable insights into how community members were engaged in different stages of waste-management initiatives.

Notably, while three case studies incorporated direct upcycling activities, the second case study (Yommarat Village) contributed alternative insights into the challenges of community engagement in waste-separation practices. This case also underscored the critical importance of clearly communicating the benefits of participation, a factor that significantly influenced motivation and engagement levels.

During the planning phase, several strategies were identified. Lesson 1 highlights the importance of clear communication of roles and responsibilities in participatory processes, echoing recommendations by DEQP (2023), which stress that transparent communication fosters trust and long-term commitment to community waste-management projects. Lesson 3 similarly aligns with Warin Chamrab Municipality (2018), which advocates for easy-to-follow guidelines in waste sorting and recycling as key factors in promoting participation. Lessons 4 and 7 both demonstrate that highlighting the tangible benefits of participation such as cost savings, income generation from recyclable sales, or environmental improvements, is key to motivating individuals. In addition, Lesson 8 (from the overall analysis) emphasizes the value of segmenting community members based on their roles to foster better collaboration and ownership of activities.

In the operational phase, the literature underscores the value of linking waste-management activities to local cultural practices. Lesson 2 demonstrates that aligning upcycling activities with community lifestyle and traditions increases participation, which is consistent with Phetthirasawat (2023) who discusses how traditional craftsmanship can be leveraged to strengthen community ownership in upcycling initiatives. Upcycling workshops were identified as a critical mechanism for enhancing community engagement and promoting creative reuse (Nimthong, Sinchao, Patthamasonti, Sirinakhon, & Faiyue, 2024). Lesson 5 points to the value of integrating design knowledge into waste-management workshops to spark creativity and expand reuse opportunities. Similarly, Lesson 6 underscores that building local capacity in design skills is essential to sustaining long-term community engagement in upcycling and waste reduction.

This chapter has identified both exemplary waste-management practices and areas for improvement across the case studies. A recurring theme is the need to more clearly integrate the 'participation in receiving benefits' step within community participation models, as observed in Yommarat Village and other settings where engagement was weaker. This reinforces Maschamadol's (2018) argument that perceived benefits are a key driver of community involvement in public initiatives (Maschamadol, 2018).

Finally, this research suggests that future studies should further investigate motivational factors that drive community participation in creative waste-management practices, such as upcycling. Understanding the interplay between cultural context, economic incentives, and community leadership could contribute to the development of more effective participatory models for small communities. Doing so would advance sustainable waste-management practices, both within Thailand and potentially in other similar socio-cultural contexts.

While this study provides insights into participatory waste management and upcycling in small Thai communities, several limitations must be acknowledged. First, the research is based on four case studies, which, while varied, may not fully represent the diversity of rural communities across Thailand or other ASEAN countries. Second, one case study (Yommarat Village) lacked upcycling activities, limiting the study's direct exploration of participatory upcycling in that specific context. Additionally, some findings were derived from secondary data sources, which may limit the depth of contextual understanding.

Future research should focus on applying these findings in a broader range of communities, including urban or peri-urban settings, to test the adaptability

Table 15.2 Summary of key lessons from the four case studies

No.	Case Study	Location	Lesson
1	Case Study 1	Tunglooknuk Sub District, Nakhon Pathom Province	**Lesson 1:** Clear communication of roles and responsibilities within the participation process. **Lesson 2:** Aligning upcycling and waste-management activities with community culture and lifestyle to enhance participation.
2	Case Study 2	Yommarat Village, Tha Nang Ngam Subdistrict, Phitsanulok Province	**Lesson 3:** Providing explicit waste-separation guidelines to encourage initial participation in waste management. **Lesson 4:** Communicating tangible benefits of participation to foster motivation and engagement.
3	Case Study 3	Waeng Yai Sub District, Khon Kaen Province	**Lesson 5:** Integrating design knowledge into workshops to enhance creative reuse and innovation.
4	Case Study 4	Wiang Phrao Sub District, Chiang Mai Province	**Lesson 6:** Capacity-building through training workshops focused on creative thinking and upcycling. **Lesson 7:** Using external incentives and recognition (e.g., awards) to stimulate and sustain community participation.

of the participation models identified. In particular, further investigation into motivational factors such as incentives, cultural drivers and perceived benefits, could help refine strategies to enhance participation in community-led upcycling projects. Developing a comprehensive participatory design framework tailored for rural upcycling initiatives would also provide practical value for policymakers and local organizations aiming to improve sustainable waste management outcomes.

Acknowledgements

We are deeply grateful to our esteemed supervisors, Dr Busayawan Lam and Dr Eujin Pei, whose exceptional guidance and unwavering support have been invaluable throughout our research journey. Their expertise, mentorship and encouragement have played a pivotal role in the successful completion of this chapter.

References

Assegaff, S., Kurniabudi, K. and Fernando, E. (2016), 'Impact of extrinsic and intrinsic motivation on knowledge sharing in virtual communities of practices', *Indonesian Journal of Electrical Engineering and Computer Science*, 1(3): 619–626.

Atthaporn, P., Kittikronwaranon, P., Tarasi, K., & Yolai, N. (2024). Solid waste management in Khon Kaen Municipality. J*ournal of MCU Ubon Review, 9*(1).

Buranasingh, N. (2017), 'Parliament', Available at: https://www.parliament.go.th/ewtadmin/ewt/parliament_parcy/ewt_dl_link.php?nid=42894 (accessed 15 July 2023).

Chainam, N. and Bouphan, P. (2022), 'The participation in waste management of people at Waeng Yai Municipality Waeng Yai District Khon Kaen Province', *KKU Research Journal*, 22(2): 224–236.

Chiangka, N. (2018). *Public participation in community waste management of Tunglooknok Subdistrict*. Bangkok: Master's Thesis Mahachulalongkornrajavidyalaya University.

DEQP (2023), 'Department of Environmental Quality Promotion', Available at: https://datacenter.deqp.go.th/knowledge/environmentdoc/12/ (accessed 10 July 2023).

Limsomboon, W., Likitpanyarat, N., & Srisawang, K. (2024). Waste management issues and needs in Thungluknok Community, Kamphaengsaen District, Nakhonpathom Province. Nakhonpathom: the 16th National Conference, Nakhon Pathom Rajabhat University.

Maschamadol, P. (2018), 'King Prajadhipok's Institute', Available at: http://www.kpi.ac.th/index.php?title=การมีส่วนร่วมของประชาชนตามรัฐธรรมนูญแห่งราชอาณาจักรไทย_พุทธศักราช_2560 (accessed 31 July 2023).

Ministry of Interior (2022), 'Ministry of Interior', Available at: http://www.chumpakpaew.go.th/fileupload/3556659235.pdf (accessed 16 July 2023).

Mokhsart, N. (2022), 'Think Forward Center', Available at: https://think.moveforwardparty.org/article/environment-and-resources/2922/ (accessed 15 July 2023).

Nimthong, P., Sinchao, W., Patthamasonti, T., Sirinakhon, C., & Faiyue, R. (2024). *A participatory waste management project at Yommarat Village, Tha Nang Ngam Subdistrict, Phitsanulok Province*. Songkhla: Faculty of Humanities and Social Sciences, Thaksin University.

Phetthirasawat, T. (2023), 'The Process of Transferring Wisdom Inheritance for Villagers' Self-Reliance in Phukhaothong Sub-District, Phranakhon Si Ayutthaya District Phranakhon Si Ayutthaya Province', *Journal of MCU Social Development*, 8(1): 341–356.

Pollution Control Department (2019), 'Community Based Solid Waste Management', Bangkok: Department of Environmental Quality Promotion Ministry of Natural Resources and Environment.

Pollution Control Department (2022), 'Pollution Control', Available at: https://thaimsw.pcd.go.th/report1.php?year=2565 (accessed 15 July 2023).

Tha Nang Ngam, S. A. (2023). The Nang Ngam Subdistrict Administrative Organization. Available at: https://www.thanangngam.go.th/news_atv.php (accessed 14 July 2023)

Thanangngam Subdistrict Administration Organization. (2024). Thanangngam Subdistrict Administration Official Website. Available at: https://www.thanangngam.go.th/home (accessed 28 November 2023).

Thungluknok Subdistrict Administration. (2024). Thungluknok Subdistrict Administrative Official Website. Available at: https://www.thungluknok.go.th/backup/homepage (accessed 17 December 2023).

UN Environment Programme (2017), 'Waste management in ASEAN countries', Available at: https://www.unep.org/resources/report/waste-management-asean-countries-summary-report (accessed 23 July 2025).

Warin Chamrab, M. (2018), 'Warin Chamrab Municipality', Available at: https://warincity.go.th/index.php?option=com_content&view=article&id=4016:2018-02-28-07-34-25&catid=121:2013-03-27-14-28-54&Itemid=414 (accessed 17 July 2023).
Wiang Phrao Municipality. (2023). *Wiang Phrao Municipality Official Website*. Available at: https://wiangphrao.go.th/index.php (accessed 21 December 2023).
Yin, R. K. (2017), *Case Study Research and Applications*, California: SAGE Publications.

Chapter 16

SUCATA QUÂNTICA: LOCAL REUSE ACTIONS IN SÃO PAULO, BRAZIL

Hamilton Ortiz

Introduction

In Latin America, there are few countries with robust industrial production (e.g., Brazil and México) despite the dominance of large agriculture, livestock and raw materials production. Latin America also suffers from a lack of efficient transportation, which implies that the destination of waste is not optimized. Goods are imported ready for use; a few years later, they are landfilled or, in the best case scenario, recycled. An alternative way to address the waste management challenges is by upcycling materials and objects in manufacturing cycles. The aim of this chapter is not to study large-scale initiatives or industries but to introduce one good small-scale case study (or best practice) in São Paulo, Brazil, to inspire and inform other upcycling practitioners (or other relevant stakeholders) to strengthen the local capacity to design a new wave of manufacturers who fully understand the risks and potentials of each material. The best practice is from Sucata Quântica – the author's initiative (or workspace) to reuse objects and materials, which nowadays acts as an educational, eco-design, and eco-fabrication movement. Sucata Quântica is a driving force behind many upcycling interventions in São Paulo and is continuously working to set up and integrate pilot reuse facilities. The evolution of the workspace between 2018 and 2023, as well as the 18 in-person workshops, provide important insights into, among other things, the motivations, enabling conditions and characteristics of the population. Based on the insights, the author suggests ways to scale up the concept of upcycling for a variety of materials, with a focus on electronic equipment.

Sucata Quântica

The main motivation for the author to start Sucata Quântica was the shocking amount of discarded materials and objects on the streets, sidewalks and dumpsters of São Paulo, Brazil, in contrast to other countries such as Colombia. Initiatives such as Unilabor in Brazil, the Reserve des Arts in France, or reuse centres such as CCRX in Chicago served as a reference and inspiration for the author. The assembly of a workspace to develop the practice of eco-design and eco-manufacturing was based on the premise of taking advantage of the recovered materials and producing the infrastructure of the workspace itself. Workbenches, shelves, boxes, lighting, sockets, and hydraulic installation of the workspace were built in large part with the reused materials. Since 2018, five different spaces have been used for Sucata Quântica: four in downtown São Paulo and one in the Ipiranga neighbourhood, initially with very limited space, but growing all the time. The last and current location is a $350m^2$ mezzanine in a building that used to be a shoe factory (Scatamacchia), which had been unoccupied and neglected for over 10 years. The workshop underwent a retrofit to accommodate the new workshop and plumbing, electricity, gates, and other basic necessities were installed. The whole process used waste materials that were independently collected, classified and self-managed. Currently, the space has facilities for metalwork, carpentry, sewing, electrical and electronics, serigraphy (screen printing), 3D printing and painting. It also maintains a permanent stock of materials and tools. Based on personal experience, the author recommends that public reuse policies should take advantage of vacant properties in the city.

Between 2018 and 2023, Sucata Quântica organized a series of hands-on upcycling workshops (n=18). During the workshops, practical interventions of creative reuse were used to teach emancipatory practices such as basic electronics to build circuits, sewing, serigraphy, sculpture, and carpentry with hand tools, amongst others (Figure 17.1). The workshops were conducted in both the private (e.g., community centres) and public spheres (e.g., parks), with all kinds of audiences in terms of age (from 10 to 70 years old), gender, nationality and social conditions. The workshops were always collaborative and *popular education* activities in which participants contributed with their educational backgrounds and experiences. Each workshop was designed with a specific objective in terms of materiality, techniques and target audience. The most common workshop was the reuse of electro-electronic equipment (EEE) for building lamps (Figure 17.2). In this case, participants learned the basics of electronics for lighting and in a three-hour workshop, were taught to build a lamp. The workshop integrated practices such as bricolage, tinkering and basic technical skills that everyone can easily learn. Most of the workshops focused on the integration of concepts of EEE repair, mechanics, lighting and sound equipment with energy and resource efficiency in mind, the use of simple, popular infrastructure to address a complex problem, and the full integration of upcycling with societal needs and urgencies, such as efficient lighting equipment, efficient ventilation equipment, infrastructure for urban farming, household recycling systems, among others.

The usual lamp-building workshop process started with selecting the base materials for the luminaire, followed by cleaning the parts. The process also involved selecting and fixing hardware and materials for the electrical circuit, assembling and testing the electrical circuit, assembling all the parts together, and testing and final finishing. Other workshops included textile upcycling for crafting tote bags (or eco bags), building microcity models from scrap materials, and others (Table 17.1). From the workshops, it was evident that most of the participants were women and children. The proposed prototypes could all be built by every

Figure 16.1 Wood/carpentry upcycling workshop participants and their outcomes.

participant regardless of their age, gender, educational background, etc. We also witnessed the participation of some children with certain psychological conditions (e.g., autism) who also managed to follow the instructions and successfully upcycled waste materials into a functional lamp and other items. Two basic circuits have proved effective for the workshops so far: one for a direct bi-voltage connection, a switch and an LED bulb, and the other with a DC source adapter, switch and LED modules/strips.

Based on the experience of establishing and maintaining Sucata Quântica for the last six years, including the organization of the 18 upcycling workshops and some courses, the author recommends the following paths to action for the future development of reuse centres (Table 17.2).

Figure 16.2 Welding in Sucata Quântica (top left); lamps produced from the workshop (top right and bottom left); and a workshop participant with the lamp (bottom right).

Table 16.1 Upcycling workshops organized by Sucata Quântica

Location	Object	Dates	Participants	Number of participants
Sesc (Commercial Social Service) – Itaquera	Construction of light fixtures	2018, 2019	General public	15 people per session
Sesc – Guarulhos	Construction of eco bags and folders	2019 (2 sessions)	General public	15 people per session
Sesc – Guarulhos	Scrap Microcities	2019 (2 sessions)	Children	30 people in total
Occupation Ouvidor 63	Construction of light fixtures	2019, 2022	General public	14 people
São Remo – Community space	Construction of light fixtures	2022	General public	16 people
Jardim Monte Kemel – Communitary space	Construction of light fixtures	2022	General public	18 people
Arts Square – Municipal Theatre	Teatralhas Quânticas	2023 (4 sessions)	General public	20 people per session, a total of 80 people
Sesc – Guarulhos	Construction of light fixtures: Ovnis	2023 (4 sessions)	General public	45 people per session, a total of 180 people

Table 16.2 Recommendations for future reuse centres

Category	Recommendations (paths to action)
Websites with data repositories	Online databases of reuse and upcycling initiatives
	Online databases of tutorials and workshops
	Short videos addressed to local authorities and key players
	Online database of resources for children's engagement
	Design handbook for upcycling centres
	Toolkits for the detailed design of reuse modules
Detailed design of facilities	Certification frameworks for health and safety (H&S) assurance
	Product storytelling guide
	Toolkit of object-oriented strategies in the design process
Detailed design of products	Tutorials and detailed documentation of product collections in Open-source Licensing
	Public policy design of massive deployment of reuse centres
Content development to engage donors and partners	Experience-based research for adaptation of reuse modules
	Business models for cooperatives

Conclusion

Sucata Quântica, the local reuse centre, and the work of upcycling workshops developed by the author as design interventions were introduced to demonstrate the feasibility of sustainable local reuse centres and actions for the long term. We witnessed that upcycling workshops are powerful ways to scale up practices within the general public, even those with limited access to technology and higher needs for basic resources. All the workshops showed an emotional impact of the finished pieces and the process of making them. From the participants' testimonies, it was clear that upcycling is not commonly practised in everyday life and that when it becomes a real experience, it encourages changes in consumption patterns and behaviours. There were also notable cases of positive impact within commonly excluded social groups, such as the elderly, children with mental health problems (mainly autism) and the unemployed. There is great potential to develop new production processes related to products and material handling. Further work is needed to improve the scalability of the processes introduced here. This work needs to be a transdisciplinary effort with a focus on skills.

Acknowledgements

The author would like to acknowledge the people of Sucata Quântica, a collective initiative founded by the author. Special acknowledgement to Maria Cecília Loschiavo, Lara Leite (USP teachers), Iara Pinheiro, Kelvin Oliveira and Pedro Tavares.

Chapter 17

Community Empowerment Through Upcycling Pre-Consumer Waste Material

Walter Chipambwa, Edlight Mutungwe and Chipo Chuma

Introduction

The apparel manufacturing sector is well known for contributing to the negative impact it has on our environment. Due to the worrisome levels of greenhouse gases (GHG) in the atmosphere, the quick depletion of natural resources like water and petroleum and the rising levels of industrial effluents, every production activity, particularly in the textile sector, must be scrutinized for its sustainability (Textile Magazine, 2020). In the European Union (EU), the textile industry alone generates around 16 million tonnes of textile waste annually (European Commission, 2017) much of which ends up in landfill or is incinerated. Accounting for the textile waste generated has also become a huge problem as supply chains have become more complex (Govindan & Hasanagic, 2018). The growth of fast fashion over the past 20 years has resulted in an excess of cheaply constructed clothing, leading to greater rates of clothing waste (McNeill & Moore, 2015; Choi et al., 2014). The textile industry has implemented several strategies to combat the issue and lessen its impact on the environment. Textile recycling, which involves both the reuse and replication of fibres from textile waste is one of these approaches and, more recently, the upcycling of textiles has been promoted as a sustainable strategy to combat this waste. Back in the fifteenth to eighteenth century, regardless of the source of their clothing, upcycling was a must as people hired seamstresses or designers to make clothes for them irrespective of their social class (Traynor, 2023). The rich would buy expensive gowns which they would repurpose as they did not throw away good camel hair and expensive laces used on their outfits. Block (2022) reveals that upcycling among the wealthy was common during the 1800s, for example the bridal gown of Alexandra, the Princess of Wales was transformed by her designer into an evening dress shortly after her marriage to Edward VII in March 1863. These high-quality gowns took a lot of time to construct, so users would upcycle or repurpose them for a specific function. The concept of upcycling was also popularized during the 1930s and 1940s, a time when families had very limited financial means and material resources due to the Second World War (Textile Magazine, 2020). The period witnessed a trend where people would recover pieces of fabric or cloth from used or unwanted clothes and remake them into something useful. With the advent of fashion shows and famous people sporting clothing made from recycled material, the use of upcycled fabrics truly took off in the 1960s (Sharma, 2019). Sung (2015) defined upcycling as an activity of creating newness or better quality from used or wasted materials while adding value. Wilson (2016) also defined upcycling as a practice of refashioning something useless into something useful without wasting any of the raw materials and without the input of non-renewable resources. In other words, upcycling is the creative technique of turning trash into an innovative piece of furniture, apparel and other products/objects. In recent years, it has grown in popularity among environmentally conscious designers and creatives. The fact that it gives worn-out garments a feeling of worth is one of the key factors contributing to its popularity. It means that consumers prefer new garments made from these fabrics rather than throwing away their old ones. Due to the high cost of new products, upcycling is a better option for those living in rural parts of developing nations to get access to items like clothing as they can use waste materials to come up with creative products.

The fundamental distinction between upcycling and recycling is that, in recycling, the material is broken down and transformed into new material, whereas in upcycling, the material is kept in its original form and used for a new purpose (Wilson, 2016). With regards to energy usage, recycling uses more energy than upcycling, as energy is usually used to disassemble the material. Fewer resources are required in the production of products when products/items are reused (eg, using virgin materials). Since materials are recycled or reutilized rather than being thrown away, less energy is spent throughout the design process. When designers reuse components from previously abandoned garments or objects, the original goods are more likely to be used to their full potential (Niinimäki, 2017). The benefits of upcycling are that it reduces the need to produce more raw materials used in manufacturing, reduces air and water pollution and the amount of greenhouse gas emissions (Satam, 2021; Caldera et al., 2022). Through creative design, upcycling can also give unique designs that can become part of trendy collections that show a high level of creativity and exclusivity in design. Upcycling benefits communities, the planet, and the users as it generates income for people while saving the planet and giving the customers a unique feel through self-expression.

Jacometti (2019) opines that the current linear model in the fashion industry does not support sustainable development goals, thus calling for a shift towards embracing the circular economy (CE). Todeschini et al. (2017) state that upcycling and recycling are important facets of the implementation of CE in the fashion industry. Recycling and reusing the fabric instead of throwing it away is what the CE in the fashion business entails. The take-make-waste approach may be substituted with a recover-recycle-reuse model using circularity in fashion. This creates value addition by converting trash into resources that can be reintroduced into the value chain. (Shirvanimoghaddam et al., 2020; Aus et al., 2021)

Before garments reach the intended consumers, a lot of waste will have been generated along the way. Approximately 15 per cent or more of fabric meant for clothing is left as waste on the cutting table (Textile Magazine, 2020). Pre-consumer textile waste comprises fabric remnants, faulty materials, and other waste generated during the manufacturing process (Haq & Alam, 2023). The cutting section in any garment manufacturing setup generates waste in the form of fabric scraps, faulty fabric, excess fabric spreading loss and all these contribute to pre-consumer textile waste. Due to the natural shapes that are utilized in creating a garment, 10–30 per cent of the cloth is removed and abandoned during the cutting process (Dimmick, 2023). This includes wastes like textile fabric swatches from the production process, end-of-roll pieces, cut and sew waste, damaged fabrics, and clothing samples. Process waste in clothing is unavoidable as it is dependent on the garment's pattern design, whereas incidental waste is regarded as avoidable. Both types of waste do contribute significantly to the pre-consumer waste generated in the apparel and fashion industries. Through upcycling, all this waste can be turned into something useful as each designer has a unique approach to manipulating the waste.

The concept of upcycling textile waste material has gained significant attention as a means of community empowerment. Upcycling at the community level can be instrumental in bringing positive change in society. One such project is the 'Stitched Up' project in Manchester, UK. It has enabled individuals to take charge of their appearance through alternative fashion whilst improving the environment and interacting with their community (Sharma, 2019). In another case in Portugal, the project 'From Granny to Trendy' is made up of women over 50 who do fashion upcycling and earn extra income from the project (Machado, 2022). Environmentally aware customers are requesting more eco-friendly items as the green movement expands in the apparel sector. Communities are also taking part in this movement as it can help to drive economic empowerment and plays a key role in employment creation. In their study, Wu et al. (2022) argue that there is a dearth of literature on community-based clothing recycling and upcycling endeavours. This project sought to explore upcycling as a practice at the community level so that it could trigger more interest from the public as the participants designed various upcycled products for use and sale.

Lately, apparel manufacturing has shifted to emerging economies as companies seek to maximize their potential on profit margins, thus posing a key issue on waste management in such countries. In most cases, these companies are enticed to relocate to these developing nations due to cheap labour, thus implying more profit since the apparel manufacturing process is labour-intensive. It then becomes very important to find ways of dealing with the waste generated in these economies as it contributes to GHG emissions. One such method is the establishment of community upcycling projects such as the 'Musoromuchena Upcycling Project' that can empower communities and, at the same time, reduce waste.

Approach

The Musoromuchena project involved women in the Makonde district in Mashonaland West, Zimbabwe. This group was made up of various age groups who were given pre-consumer waste fabrics to develop products they deemed appropriate. The project was a result of action research that established a community need in the Musoromuchena communal area. It collaborated with a non-governmental organization (NGO), Musoromuchena Skills for Life Centre. The Centre had a large piece of land that could accommodate a sewing room, farming plot and laboratories for other skills-learning activities. Over sixty women participated in the pre-consumer waste action research under the guidance of a trainer and the research team. Some of the women who lacked basic sewing skills were trained at the centre to make sure they could handle and operate sewing machines. The project was more vibrant during the farming off-season period, providing the women with enough time to interact and discuss how they could design useful products from the waste materials. Indigenous knowledge systems (IKS) were a component of the project that guided the women in choosing products to work on using the waste materials. The women managed to showcase the products they designed in an exhibition at the centre in July 2022. Figure 18.1 shows the materials used by the women in the upcycling project.

Each exhibitor was allowed a time slot to describe the process they used in coming up with the product and the functionality aspect, as well as to state any challenges they faced during the process. The exhibitors highlighted the benefits they had accrued due to the upcycling project. Figure 18.2 shows the women sorting and selecting the waste materials as the women before sharing the bundles. The pieces were selected based on fabric type, colour and the pattern style of the offcuts. This ensured that everyone get similar pieces wherever possible.

Figure 17.1 Textile fabric waste.

Figure 17.2 Sorting and selection process.

Outcomes and Challenges

The exhibition allowed everyone to show the products they had designed and produced using waste materials. All the materials were sourced from a clothing manufacturing entity affiliated with Chinhoyi University of Technology. The centre fully sponsored the exhibition event.

Social Benefits of the Project

The exhibition showcased several products, including bags, bed covers, ropes, skirts, shorts, rugs and many other items. The project managed to increase social interaction among the women in the area as they met once or twice a week to share ideas about products that could be designed using the waste materials. More than 70 per cent of participants mentioned that the project had led to a positive improvement in their livelihoods. One participant mentioned that the engagement and teachings they got from other members and facilitators helped them to deal with issues of abuse in their home. Through the project, some participants also mentioned that it managed to keep their minds active as they explored ways of manipulating the waste so that they could reap more benefits from it. One of the participants mentioned, 'The project keeps my mind busy as I have to come up with creative ideas of sewing or knitting the various types of fabric waste, and that is healthy for me, unlike spending the whole day idle at home' (Participant 3). The participants also gained skills, such as sewing and knitting, through this project. It improved the awareness of the local community on issues with sustainability and, at the same time, boosted their creativity.

Economic Benefits of the Project

Some participants generated income by upcycling the provided waste fabric and selling their products to other people within the community. As shown in Figure 18.3, one participant made bed covers, sold them and used the money to buy food and household utensils for her family.

In a similar case, another participant mentioned that she managed to make ropes to use at home for activities such as farming and fetching water from the well. She went on to state that the husband initially had a negative view of the project but, after seeing her making useful products at home, he became supportive of her efforts. Another participant also exhibited small bags that could be used at home or at school to carry books, as shown in Figure 18.4.

Summary of Challenges Faced

Despite the project providing numerous benefits highlighted by women who played an active role, several problems were also highlighted by the participants. The major problem was the equipment used for the sewing process as they only had access to hand-sewing machines. One participant pointed out that they would be happy to purchase more industrial sewing machines and other specialized machines such as buttonholes and overlocking machines. Another participant stated, 'The quality of our products is affected by the type of machine we use; using better industrial sewing machines can result in marked improvement of our products as we sometimes end up stitching using our hands' (Participant 6).

It was also emphasized that participants could not source or access matching sewing threads for the fabric pieces, posing a challenge to the construction process. The participants also mentioned that it was difficult to conveniently meet as one group for training sessions due to family commitments and the demands of other

Figure 17.3 Bedcovers made from upcycling.

Figure 17.4 Denim upcycled bag.

activities in their homes and communities. In the area, Thursday is apparently a sacred day, and no one is expected to do any farming activities. Therefore, this was the day when the women got time to come to the centre for their project. However, they said it was not enough. Some women had to walk long distances to the centre, which affected their attendance and they missed out on learning opportunities. The selection process was another challenge as the women had little knowledge of the various fibre compositions of the materials and, in some cases, failed to sort the related materials correctly. Sourcing the pre-waste from one place can be a challenge in terms of getting a variety of materials. As one participant mentioned, 'We need to have a variety of fabric pieces in terms of quality, colour, and size so that we can widen the product ranges we make.' (Participant 2). Another participant also echoed a similar sentiment as she stated that they would prefer fabric pieces from blankets so that they could make winter coats, jackets and blankets. The frequency of supply of waste fabric materials was also highlighted: the participants mentioned that they could do more if they had more and better materials

Discussion

This action research demonstrated that the upcycling of pre-consumer textile waste could bring numerous benefits to the community, companies and the environment. In this study, the community exhibited products that were made from waste fabric; some managed to earn income from selling their products. There is a need to improve the design skills of the women participating in the upcycling project, as most exhibited a lack of sewing and design skills. The lack of equipment, such as specialized machines was mentioned and, as such, sourcing for such equipment from other clothing manufacturers could be a solution. It should be noted that upcycling is not just the responsibility of companies and consumers having government policies that encourage manufacturers to act and promote CE initiatives is equally important. To be a good designer who can upcycle, one needs to be educated or trained in how they can use design thinking to come up with unique products. In this project, the training was facilitated by one individual in various areas. Having a designer who would come to teach the women how to develop new products creatively could also be explored as the project is ongoing. Engaging various facilitators can be instrumental in improving the output of the upcycling project. Involving different facilitators can contribute to the progress of the upcycling project. This community upcycling initiative has influenced the livelihoods of participants, consistent with findings reported by Machado (2022) and Wu et al. (2022), who noted that such projects can affect participants' outcomes.

The success of upcycling is contingent upon the availability of waste materials, a factor that proved to be a significant obstacle for the participants. The project demonstrated that it had long-term potential provided that the appropriate support structures were established. One such key intervention is the engagement of a full-time fashion instructor or designer who, through training, can equip the women to upscale their design capabilities. The project has improved people's awareness of sustainable fashion and, at the same time, has enhanced community engagement. Networking with apparel manufacturers can improve community upcycling projects by allowing better waste sorting at the company level. Using waste from the apparel industry helps reduce landfill use, conserve resources and lower emissions by decreasing the water and energy required. Governments can encourage sustainable practices using tax incentives, while manufacturers should also back community efforts to advance the CE. The government can also come up with mechanisms to promote sustainable behaviour amongst manufacturers by offering tax rebates for those who score highly in terms of their sustainability activities. Manufacturers need to support community-based sustainability initiatives as they promote the circular economy model.

Conclusion

Upcycling pre-consumer waste can promote sustainability initiatives in developing economies and, at the same time, help to build communities. Community engagements through upcycling can also promote awareness of the devastating effect of waste on the environment. Upcycling knowledge and training can empower communities to design craftwork products and other valuable household items so that people can earn money. As a CE ingredient, promoting upcycling at a community level can be instrumental in supporting the CE drive. By upcycling pre-consumer textile waste materials, communities can turn what might have been considered useless or discarded items into valuable products. Upcycling allows the sharing of resources between businesses and the communities in which they operate, thus fostering positive relationships. Implementation of upcycling at the community level results in psychological and sociological advantages including a sense of empowerment, relaxation and a stronger connection to society. In conclusion, upcycling pre-consumer textile waste products is an excellent approach to cutting waste, saving resources and empowering communities.

Acknowledgements

The Musoromuchena Skills for Life Centre in Mashonaland West, Zimbabwe supported this work.

References

Aus, R., Moora, H., Vihma, M., Unt, R., Kiisa, M. and Kapur, S. (2021), 'Designing for circular fashion: Integrating upcycling into conventional garment manufacturing processes', *Fashion and Textiles*, 8: 1–18.

Block, E. L. (2022), 'The Richest Fashionistas Used to Recycle Clothes as a Matter of Habit. What Happened?' Slate. Available at: https://slate.com/human-interest/2022/01/history-of-recycling-in-fashion-how-nineteenth-century-fashionistas-reworked-upcycled-and-handed-down.html.

Caldera, S., Jayasinghe, R., Desha, C., Dawes, L. and Ferguson, S. (2022), 'Evaluating barriers, enablers and opportunities for closing the loop through 'waste upcycling': A systematic literature review', *Journal of Sustainable Development of Energy, Water and Environment Systems*, 10(1): 1–20.

Choi, T.M., Hui, C.L., Liu, N., NG, S.F. and Yu, Y. (2014), 'Fast fashion sales forecasting with limited data and time', *Decision Support Systems*. 59: 84–92.

Dimmick, M. (2023), 'Intro to textile waste: Pre-consumer waste.' Available at: https://www.maggiedimmick.com/ethelstudio/2023/10/2/intro-to-textile-waste-pre-consumer.

European Union (2017), 'Mapping Clothing Impact in Europe: The Environmental Cost.' Available at: http://www.ecap.eu.com/wp-content/uploads/2018/07/Mapping-clothing-impacts-in-Europe.pdf.

Govindan, K. and Hasanagic, M. (2018), 'A systematic review on drivers, barriers, and practices towards circular economy: a supply chain perspective', *International Journal of Production Research*, 56(1–2): 278–311.

Haq, U. N., & Alam, S. M. R. (2023), 'Implementing circular economy principles in the apparel production process: Reusing pre-consumer waste for sustainability of environment and economy', Cleaner Waste Systems, 6, 100108. Available at: https://doi.org/10.1016/j.clwas.2023.100108.

Jacometti, V. (2019), 'Circular economy and waste in the fashion industry', *Laws*, 8(4): 27.

Machado, D., Ferreira, M., Carvalho, A., Braga, A. and Braga, V. (2022), 'Fashion upcycling and social impact: A case study', *Proceedings of the 17th European Conference on Innovation and Entrepreneurship*.

McNeill, L. and Moore, R. (2015), 'Sustainable fashion consumption and the fast fashion conundrum: Fashionable consumers and attitudes to sustainability in clothing choice', *International Journal Consumer Studies*, 39: 212–222.

Niinimäki, K. (2017), 'Fashion in a circular economy', in C.E. Henninger, P.J. Alevizou, H. Goworek, D. Ryding (eds), *Sustainability in Fashion*, 151–169, Cham: Palgrave Macmillan.

Satam, S. (2021), 'Understanding Upcycling: Being Conscious of Your Footprint'. Available at: https://www.organicamrit.com/blog/Understanding-Upcycling-Being-Conscious-of-Your-Footprint.

Sharma, A. (2019), 'Upcycled clothes, sustainable fashion, educating communities: The story of Stitched Up', *Ideas Alliance*, Available at: https://ideas-alliance.org.uk/hub/2019/10/17/upcycled-clothes-sustainable-fashion-educating-communities-the-story-of-stitched-up/ (accessed 25 July 2023).

Shirvanimoghaddam, K., Motamed, B., Ramakrishna, S. and Naebe, M. (2020), 'Death by waste: Fashion and textile circular economy case', *Science of the Total Environment*, 718: 137317.

Sung, K. (2015), 'A review on upcycling: Current body of literature, knowledge gaps and a way forward', *International Conference on Environmental, Cultural, Economic and Social Sustainability*, 17(4): 28–40.

Textile Magazine (2020), 'Upcycling of Pre-Consumer Textiles Waste', Available at: https://textilevaluechain.in/in-depth-analysis/articles/textile-articles/upcycling-of-pre-consumer-textiles-waste (accessed 25 July 2023).

Todeschini, B.V., Cortimiglia, M.N., Callegaro-de-Menezes, D. and Ghezzi, A. (2017), 'Innovative and sustainable business models in the fashion industry: Entrepreneurial drivers, opportunities, and challenges', *Business Horizon*, 60(6): 759–770.

Traynor, S. (2023), 'Upcycling: A Brief History', Slow Fashion USA. Available at: https://www.slowfashionusa.com/blog/upcycling-a-brief-history.

Wilson, M. (2016), 'When creative consumers go green: Understanding consumer upcycling', *Journal of Product & Brand Management*, 25(4): 394–399.

Wu, D., Zhuang, M., Zhang, X. and Zhao, Y. (2022), 'Towards circular fashion: Design for community-based clothing reuse and upcycling services under a social innovation perspective', *Sustainability*, 15(1): 262.

Part V

Design Education

Chapter 18

INCORPORATING UPCYCLING INTO THE INDUSTRIAL DESIGN PROGRAM: THE CASE OF THE UNIVERSITY OF BOTSWANA

Richie Moalosi, Polokano Sekonopo, Yaone Rapitsenyane and Paulson Letsholo

Introduction

The world is slowly transiting to new economic paradigms that intend to implement the circular economy's guiding principles and seek to provide the conditions for sustainable development. People's awareness of sustainability and the value of conservation and upcycling has increased because of the global environmental catastrophe. The Sustainable Development Goals (SDGs) are a set of objectives set by the United Nations to foster sustainable development to improve human lives and protect the environment (United Nations, 2015). The SDGs urge communities worldwide to learn about and participate in sustainable initiatives such as upcycling with more informed actions. The University of Botswana has incorporated sustainable upcycling into its design programme to demonstrate its unwavering commitment to contributing to efforts to combat climate change. The design programme is based on designing products, services, and systems. The best option to address the pressing issues of waste disposal is to introduce upcycling to students early in their academic careers as one of the sustainability measures (Stahel, 2016). While there is growing interest in incorporating upcycling into the Industrial Design programme at the University of Botswana, there appears to be a research gap regarding the long-term impact of upcycling education on students and the broader design community.

Incorporating upcycling into the Industrial Design programme at the University of Botswana is a strategic decision that not only aligns with environmental imperatives but also empowers students by providing them with a versatile skill set, local relevance and a global perspective. This initiative serves as an example of the university's commitment to fostering responsible, innovative and community-engaged designers who can address the challenges of today's and future world. This study was based on a case study conducted with industrial design students at the University of Botswana. Upcycling was integrated into the programme through a project-based learning approach to equip students with knowledge and skills about sustainable development principles. The aim was to expose students to different ways of conserving resources, reducing waste, assessing environmental impact, posing a creativity and innovation challenge of using waste materials, inculcating social and cultural values on upcycling and empowering students to be ambassadors of the upcycling concept. This approach is aligned with Shuhailo et al. (2022), who argue that higher education should promote the ideas of the circular economy and the development of ethical standards for the reuse of products. By encouraging students to recognize the advantages of upcycling, they will be more aware of their environment and transform waste materials or unwanted products into new, functional and often higher-value items needed in the circular economy (Maaruf, 2021).

Circular Economy Practices

The earth can no longer sustain the level of resource demand that humanity needs; here is thus a significant environmental deficit in the context of rapid economic growth (Wu & Bai, 2022). The world is facing a dilemma due to a linear economy based on perpetual growth and consumption of finite resources (Abrishami et al., 2021). Resource extraction will remain intense and in the future to meet humanity's needs (Anastasiades et al., 2020; Arif et al., 2012). This has given rise to various global challenges, including climate change, environmental degradation, energy scarcity and other threats to humanity and security (Ali et al., 2023). There is an increased understanding of the complex relationship between environmental concerns and social development in our rapidly changing world. This realization has amplified the significance of sustainable development in the circular economy (Khan et al., 2023). A circular economy is an economy that produces zero waste (Afteni et al., 2022). In a circular economy, resources are replenished. Processes, including maintenance, reuse, refurbishment, remanufacturing, recycling and upcycling, ensure that materials and products remain in circulation (Ellen MacArthur Foundation, 2023). The circular economy is based on the principles of elimination of pollution and waste, circulation of products and materials for a longer time and the regeneration of nature.

It is, therefore, imperative that we transition from the traditional linear model to a circular model to ensure the long-term sustainability of our planet (Ellen MacArthur Foundation, 2023). The linear economic model focuses on products which are produced, used and then thrown away as waste (Afteni et al., 2023). It is based on taking raw materials from the earth, using these to make products to sell and, once these products have reached their end-of-life, disposing of them. Abrishami et al. (2021) argue that, to address these challenges simultaneously, we need a radical approach to production, consumption, disposal of waste and living. Dealing with waste sustainably is crucial for reducing the environmental impact of human activities. The circular economy model focuses on designing products for durability, repairability and recyclability. It tries to prolong the time that materials are used in the economy. Currently, the following are some of the sustainable ways to manage and handle waste: reduce, reuse, recycle, composting, upcycling, waste-to-energy, extended producer responsibility, community recycling programmes, e-waste recycling, zero waste initiatives, government policies and regulations and circular economy practices. By adopting these sustainable waste management practices, individuals, communities, and industries can contribute to a cleaner environment, reduce resource depletion and create a more sustainable future (Maaruf et al., 2021). This chapter focuses on upcycling as a promising practice towards sustainable development and circular economy.

Incorporating upcycling into industrial design education not only equips students with practical skills but also instils a sustainability mindset. This prepares students to play an essential role in advancing the circular economy, where resources are conserved, waste is minimized and sustainable practices become integral to the design and production of goods. As the circular economy gains prominence globally, upcycling remains a valuable tool in achieving its goals and driving positive change.

The Importance of Upcycling in the Circular Economy

Upcycling provides novel strategies for resource efficiency and waste management and reduces solid waste and the industrial energy consumption involved in manufacturing and processing virgin materials. It is a promising alternative to reduce energy and material use and engender sustainable production and consumption. Upcycling disrupts unsustainable production and consumption patterns by combining recycling and upgrading (Wegener, 2016). Products gain value through upgrading and recycling encourages the reuse of discarded materials. Sung and Abuzeinab (2022) argue that upcycling refers to material processes and innovations using waste components, products and materials to create products of higher quality or added value than compositional elements. Therefore, upcycling offers an option for sustainable consumption by taking disposed-of products or waste into a new production and value-creation cycle. It counters the argument that an object has no value once disposed of and it must be destroyed before re-entering a new circle of production and value creation (Wegener, 2016). Adopting upcycling innovation plays a significant role in the quest for sustainable development. The goal of the circular economy is to rethink waste as a useful resource with the ability to start a new value generation chain during its existence rather than as trash (Moalosi & Sung, 2022). The circular economy promotes reducing waste and pollution, extending the useful lives of products and resources, and restoring natural systems (Bofylatos, 2022). Upcycling offers a sustainable alternative to conventional manufacturing and consumption patterns, contributing to a circular economy and a greener future. It is a significant practice that combines environmental, ethical, economic and socio-cultural benefits:

(a) *Waste reduction*: Upcycling assists in reducing the amount of waste sent to landfills or incineration facilities. By repurposing and transforming discarded materials, upcycling prevents them from ending up in the waste stream, contributing to resource depletion and environmental pollution (Bofylatos, 2022).
(b) *Conservation of resources*: Upcycling decreases the demand for new raw materials by using existing ones. It promotes resource conservation and reduces the need to extract, process and manufacture virgin materials, which helps sustain natural resources such as energy, water and forests.
(c) *Environmental impact*: Traditional manufacturing processes frequently involve substantial energy consumption and greenhouse gas emissions. Upcycling requires less energy and produces fewer emissions than creating products from scratch. By lowering the environmental impact of production, upcycling contributes to mitigating climate change and protecting ecosystems.
(d) *Creativity and innovation*: Upcycling encourages creative thinking and innovation by challenging people to find new uses and applications for discarded materials. It nurtures a mindset of resourcefulness and problem-solving, promoting unique and imaginative approaches to design and manufacturing (Duksson, 2021; Bridgens et al., 2018).
(e) *Economic opportunities*: Upcycling can create economic opportunities in local communities. It can inspire entrepreneurship and small-scale businesses centred around transforming waste materials into marketable products (Wegener, 2016). This generates income and contributes to job creation and economic development (Martín et al., 2022).
(f) *Social and cultural value*: Upcycling can have social and cultural benefits by promoting sustainable practices and raising awareness about consumption patterns and waste management (Martín et al., 2022). It can advance a sense of cooperation, community engagement and environmental responsibility, leading to a more sustainable and conscious society (Bridgens et al., 2018).
(g) *Individual empowerment*: Upcycling enables individuals to reduce waste and make a positive environmental impact. It also enables individuals to actively participate in sustainable practices by reimagining and repurposing items in their residences or societies (Martín et al., 2022). This sense of empowerment can lead to a greater sense of well-being and personal fulfilment.

Methods

A case study was conducted with 22 third-year Bachelor of Design (Industrial Design) students at the University of Botswana. All students doing the module 'IBC322 Integrated Product Design' participated in the project as part of their programme fulfilment. A case study was selected for this study because it is an empirical investigation that examines a phenomenon in its actual setting (Yin,

2014). The methodology promotes the deconstruction and subsequent rebuilding of numerous phenomena and enables the researcher to investigate persons or organizations through interventions, relationships communities, or programmes (Priya, 2021; Yin, 2014). A project-based learning approach was adopted to build students' creative capacity to work through complex or ill-structured problems, commonly in small teams. The fundamental tenet of project-based learning is that students learn to apply new knowledge in a problem-solving environment while engaging with and considering real-world challenges (Funke, 2022). The educator is a facilitator who enables students to construct knowledge through exploration and experimentation.

Students were introduced to a series of lectures and group discussions on upcycling, which included some of the following topics: an introduction to upcycling, materials and tools, upcycling techniques, design inspiration, upcycling projects, upcycling and sustainability, the business of upcycling and the future of upcycling. In parallel with the lecture series, students were given an open design brief to design a product of their choice using waste materials. They followed the following six steps:

1. Identify and collect the waste materials they could use for their project.
2. Clean and sort the waste materials.
3. Design and plan.
4. Deconstruct and prepare materials.
5. Upcycle and assemble.
6. Apply finishing touches.

The project was completed in 14 weeks, and the participants spent four hours a week on it. Students worked in collaborative groups of three members and used materials such as old tyres, bottles, scrap metals, wood off-cuts, cardboard boxes, plastics, etc. The aim was to expose students to upcycling and align with the goals of fostering environmentally responsible citizens equipped with critical thinking skills, creativity and a sense of agency to address our world's complex sustainability challenges. The data was collected from the students' portfolios and was analysed through thematic and visual analysis. The following sections present a selection of the projects undertaken by the students.

Findings

Since this was the student's first exposure to upcycling, the focus was on exposing them to upcycling rather than emphasizing the quality of the products produced. Figure 20.1 (left) shows a reception chair made from waste cardboard boxes. After making the chair's structure, upholstery was applied to enhance its aesthetic. This product shows that waste can be made into products of added value; this is a valuable message that should be shared with other students. Figure 20.1 (right) shows students experimenting with developing a machine that melts plastics into

Figure 18.1 Reception chair – made from cardboard boxes (left); melted plastic waste (right).

Figure 18.2 The coffee table is made from old tyres (left); the coffee table is made from wood pallets (right).

blocks, which can then be used to make tops for furniture or workstations. As such, upcycling has the potential to boost local economies by stimulating entrepreneurial activity. Small-scale companies can focus on turning waste materials into valuable and innovative products.

Figure 20.2 (left) shows a coffee table made from an old tyre, wood, and plastic off-cuts. Figure 20.2 (right) shows a coffee table made from waste wood pallets. The products exhibit a resourceful and innovative problem-solving mindset, encouraging original and creative ways to design and manufacture items from waste materials.

Figure 20.3 (left) shows a waste glass make-up organizer, while Figure 20.3 (right) shows a drinking set made from waste glass bottles and wood pallets. The

Figure 18.3 Make-up organizer (left); a set of drinking glasses (right).

glass bottles have been repurposed. Students displayed the capacity to design everyday products and add value to delicate glass bottles.

The projects in this study show that students deeply considered upcycling waste to produce products with added value. Upon analysis of the projects designed by the students, it became evident that they had acquired knowledge about environmental awareness, had participated in a hands-on learning experience, had developed a sustainable mindset and had been encouraged to express creativity and innovation. Moreover, the students were empowered to take positive action to safeguard the environment and humanity. The fact that students had to look for waste materials to design and develop their projects instilled a sense of resourcefulness, ethical consumerism and community engagement and developed a global environmental perspective on sustainable development. On a critical reflection from the interview findings, the projects have shown that teaching students about upcycling offers numerous educational, practical and ethical benefits. Incorporating upcycling into the industrial design curriculum offers the following valuable lessons to students.

1. *Environmental awareness*: Educating students about upcycling has heightened their environmental awareness, specifically regarding matters such as waste production, pollution and resource depletion. This instruction has enabled them to grasp the consequences of their consumption patterns and prompted them to adopt sustainable practices. The value of environmental education lies in its ability to prevent harm to the environment and serve as the optimal solution to such issues.
2. *Hands-on learning*: Upcycling projects offer a unique opportunity for students to engage in hands-on learning that fosters creativity and problem-solving skills. By transforming discarded items into valuable products, students were

able to apply critical thinking, practical knowledge and design skills. For example, students created specific products from reclaimed wood and other waste materials. These projects not only teach students valuable skills but also promote sustainability by reducing waste and showcasing the potential of upcycling.

3. *Sustainable mindset*: Introducing the concept of upcycling to students at an early age may instil a sustainable mindset that can have a far-reaching impact. By going beyond the linear 'use and discard' model and understanding the value of repurposing and reusing materials, students were better equipped to appreciate the importance of sustainability. When students learn about upcycling, they are more likely to continue this practice as adults, which helps foster a culture of sustainability that benefits both people and the environment in the long run. Ultimately, this contributes to a culture of sustainability that has a lasting positive impact on society and the planet.
4. *Creativity and innovation*: Upcycling is an excellent way to foster creativity and innovation, as it challenges individuals to think outside the box when repurposing materials. This results in a stimulating environment where imagination thrives, and design and problem-solving skills are improved. By adopting upcycling practices, students learn to approach problem-solving more innovatively and sustainably while also gaining a sense of satisfaction from creating something new out of something old.
5. *Empowerment*: Students who learned about upcycling felt more equipped to make fundamental changes in the world. Decreasing and/or repurposing garbage on a regular basis and making thoughtful decisions helped students to find solutions for the environment.
6. *Resource Management*: Students learnt ingenuity and economical material use through upcycling. In a world where resources are limited, students gain vital skills such as maximizing the value of already-existing resources.
7. *Entrepreneurship opportunities*: By comprehending the concept of upcycling, one can ignite eco-entrepreneurial inspiration and create/craft innovative products. Students may identify promising business prospects in the production of upcycled merchandise, which contributes to economic progress and advocates for sustainability. Eco-entrepreneurship integrates a variety of solutions with social, economic and environmental benefits for businesses, such as waste minimization, generating green jobs and introducing environmentally friendly alternatives to the marketplace.
8. *Ethical consumerism*: Teaching upcycling is an effective way to educate students about ethical consumerism, which involves considering the environmental and social implications of one's purchasing decisions. By learning how to repurpose and reuse materials that would otherwise go to waste, students gained a better understanding of the importance of sustainability and the role they can play in reducing waste and conserving resources. Additionally, discussions about ethical consumerism can help students to become more critical and thoughtful shoppers, evaluating the impact of their choices on the environment and society and making more

informed decisions as responsible consumers. Incorporating lessons about upcycling and ethical consumerism into the curriculum can help promote a more sustainable and socially responsible mindset among students.

9. *Community engagement*: Throughout the project, the students actively collaborated and shared their innovative ideas. This was instrumental in fostering a strong sense of community both within and beyond the classroom, as the students partnered with their peers, educators, and local organizations to advance sustainability initiatives. It is worth noting that each project was completed by a team of three students to maximize community engagement.
10. *Developing a global perspective*: The inclusion of upcycling as part of the curriculum has facilitated a broadening of students' perspectives on global environmental challenges. With this knowledge, students are better equipped to evaluate how their actions, even those on a small scale, can impact more significant problems such as waste management, biodiversity loss and climate change. The ability to connect their actions with these more significant issues has resulted in a more comprehensive understanding of environmental concerns. It has provided students with a sense of agency to make a positive impact on the world around them.

Overall, incorporating upcycling into the industrial design curriculum equipped students with practical skills and nurtured their sense of responsibility as stewards of the planet. It actively empowered them to build a more sustainable future in the circular economy. The results show that incorporating upcycling into the industrial design programme has achieved its objective of raising awareness, viewing waste as a valuable resource, and contributing positively to environmental challenges facing the global economy.

Discussion

Incorporating upcycling into industrial design programmes has become increasingly significant today as we strive to find sustainable solutions for waste management and resource conservation. The study highlights the significance of incorporating sustainability practices into higher education curricula, specifically within design education. The University of Botswana's initiative to integrate upcycling principles into the Industrial Design programme is a strategic approach to align with environmental imperatives and empower students by giving them a versatile skill set, local relevance and a global perspective. This study is an empirical investigation that examines a current phenomenon in depth and within its actual setting, particularly when the distinctions between phenomenon and setting are blurred. One of the main advantages of incorporating upcycling into the curriculum is that it encourages students to think creatively and innovatively about how to repurpose waste materials (Duksson, 2021). By teaching students how to transform discarded items into new and valuable products, they can develop a deeper understanding of sustainable design principles and contribute to a more environmentally friendly future. Furthermore, integrating upcycling into the industrial design programme can also help students develop practical skills in material selection, manufacturing processes, and product development (Herman et al., 2018). The hands-on experience can prepare them for careers in the design industry and make them more competitive in the green job market.

The study underscores the importance of fostering creativity, environmental awareness and practical abilities in design education (Suwaed, 2022; Ali et al., 2023). It reveals the transformative potential of integrating upcycling into the curriculum, fostering a generation of designers capable of envisioning innovative, environmentally responsible solutions (Zunaidah et al., 2023). Moreover, the study provides insights into the sustained benefits and challenges associated with integrating sustainability into industrial design programmes. For example, Jayasinghe (2020) argues that eco-entrepreneurship enables the development and introduction of eco-friendly solutions to the market. For example, we discovered that a company in Senegal makes furniture from melted plastic bottle waste. This was excellent motivation for the students because it showed that they could develop enterprises and produce furniture pieces which saved the environment from damage. However, there are also challenges associated with incorporating upcycling into the curriculum. For example, students may face limitations in terms of resources and access to materials for upcycling their projects. Moreover, there may be resistance from traditional design practices and a lack of awareness among students and faculty about the benefits of upcycling (Barratt et al., 2013).

The study findings can be used as a valuable model for other educational institutions seeking to infuse sustainability, especially upcycling, into their design programmes, contributing to a more conscious and ecologically sustainable future (Haug & Busch, 2016; Herman et al., 2018). Future research should explore the long-term impact of upcycling education on students and the broader design community. Most existing studies (Aus et al., 2021; Ali et al., 2013; Richardson, 2010) tend to focus on short-term outcomes, such as immediate skill development and immediate design projects. To address this gap, future research should explore the lasting effects of upcycling education. By addressing the long-term impact of upcycling education, research can provide insights into the sustained benefits and challenges associated with integrating sustainability into industrial design programmes. This information is crucial for programme enhancement and the continued evolution of design education in the context of global sustainability imperatives.

Conclusion

This chapter discussed the integration of upcycling principles into the Industrial Design programme at the University of Botswana, aiming to equip students with the knowledge, skills and mindset necessary to address contemporary environmental challenges. It used a project-based learning approach to expose students to different ways of conserving resources, reducing waste, assessing environmental impact and empowering them to be ambassadors of the upcycling concept. The study highlights the transformative potential of integrating upcycling into the curriculum, fostering a generation of designers capable of envisioning innovative, environmentally responsible solutions. The University of Botswana's experience serves as a valuable model for other educational institutions seeking to infuse sustainability into their design programmes, contributing to a more conscious and ecologically sustainable future.

Incorporating upcycling into the Industrial Design programme represents a significant stride towards nurturing sustainable design practices and responsible consumerism among students at the University of Botswana. This case study highlights the multifaceted benefits of introducing upcycling into higher education curricula, particularly design education. The fusion of creativity, environmental consciousness and practical skills in upcycling projects equipped students with the technical proficiency required in their field and a profound understanding of the ecological implications of their work. It underscores the university's commitment to producing well-rounded, conscientious designers capable of contributing positively to their communities and the environment. The programme exemplifies the potential for academia to shape a generation of designers who will pioneer sustainable solutions, challenge the conventional and re-imagine products and systems from an ethical and environmentally conscious perspective.

As the global community grapples with the pressing challenges of resource depletion and environmental degradation, educational initiatives such as that undertaken by the University of Botswana offer hope. By integrating upcycling into the Industrial Design programme, the university imparts knowledge and skills and catalyses a mindset shift that transcends classroom walls. This paradigm shift towards sustainability and responsible design has the power to influence industries, societies and, ultimately, the trajectory of our planet. Thus, incorporating upcycling in design education is not just a practical pedagogical approach but a profound investment in a greener, more innovative, more conscious future.

References

Abrishami, S. and Martín-Durán, R. (2021), 'BIM and DfMA: A paradigm of new opportunities', *Sustainability*, 13(17): 9591.

Afteni, C., Păunoiu, V. and Afteni, M. (2022), 'Study on theR, *Annals of "Dunarea de Jos" University of Galati, Fascicle V, Technologies in machine building*, 39(1): 49–55.

Ali, K., Kausar, N. and Amir, M. (2023), 'Impact of pollution prevention strategies on environment sustainability: Role of environmental management accounting and environmental proactivity', *Journal of Environmental Science & Pollution Research*, 30(38): 88891–88904.

Ali, M., Abduh, A., Mahmud, R. and Dunakhir, S. (2023), 'Raising students' awareness of environmental education issues', *Indonesian Journal of Educational Research and Review*, 6(1) 1–8.

Anastasiades, K., Blom, J., Buyle, M. and Audenaert, A. (2020), 'Translating the circular economy to bridge construction: Lessons learnt from a critical literature review', *Renewable Sustainability Energy Review*, 117: 109522.

Arif, M., Goulding, J. and Pour Rahimian, F. (2012), 'Promoting off-site construction: Future challenges and opportunities', *Journal of Architecture Engineering*, 18: 75–78.
Aus, R., Moora, H., Vihma, M., Unt, R., Kiisa, M. and Kapur, S. (2021), 'Designing for circular fashion: Integrating upcycling into conventional garment manufacturing processes', *Fashion and Textiles*, 8(34) 1–18.
Barratt, R., Thagard, P. and Montes, L. (2013), 'Using case studies to build the knowledge base in sustainable design', *Design Studies*, 34(1): 23–33.
Bridgens, B., Powell, M., Farmer, G., Walsh, C., Reed, E., Royapoor, M., Gosling, P., Hall, J. and Heidrich, O. (2018), 'Creative upcycling: Reconnecting people, materials, and place through making', *Journal of Cleaner Production*, 189: 145–154.
Duksson, L. (2021), 'Innovative design for upcycling waste materials – Take campus cultural and creative souvenir design as a sample', *Art and Performance Letters*, 2(6): 13–19.
Ellen MacArthur Foundation (2023), 'What is a circular economy?' Available at: https://www.ellenmacarthurfoundation.org/topics/circular-economy-introduction/overview.
Funke, J.O. (2022), 'Literature review of project-based learning', *Journal of Educational Research and Policies*, 4(7): 110–114.
Haug, A. and Busch, J. (2016), 'Towards an ethical fashion framework', *Fashion Theory*, 20: 317–339.
Herman, K., Sbarcea, M. and Panagopoulos, T. (2018), 'Creating green space sustainability through low-budget and upcycling strategies', *Sustainability*, 10(6): 1857.
Jayasinghe, R., Liyanage, N. and Baillie, C.A. (2020), 'Sustainable waste management through eco-entrepreneurship: An empirical study of waste upcycling eco-enterprises in Sri Lanka', *Journal of Material Cycles and Waste Management*, 23(2): 557–565.
Khan, A., Jie, Z., Wang, J., Nepal, J., Ullah, N., Zhao, Z.Y., Wang, P.Y., Ahmad, W., Khan, A., Wang, W., Li, M.Y., Zhang, W., Elsheikh, M. S. and Xiong, Y.C. (2023), 'Ecological risks of microplastics contamination with green solutions and future perspectives', *Journal of the Science of the Total Environment*, 899: 165688.
Maaruf, S.Z., Wan Abdul Hamid, W.N. and Sabri, M.F. (2021), 'Upcycling junk art and craft module to nurture creativity of children', *Malaysian Journal of Sustainable Environment*, 8(2): 55–70.
Martín, J. M. M., Martos, M. J. R., Pachón, A. M. and Martínez, S. C. (2022), 'Exploring what motivates the development of upcycling practices. The making of values and principles', *Ecological Questions*, 33(4): 141–152.
Moalosi, R. and Sung, K. (2022), 'Promoting upcycling through an international research network', *Proceedings of the International Online Conference on Reuse, Recycling, Upcycling, Sustainable Waste Management and Circular Economy.*
Nawwar Shukriah Ali, N. S., Khairuddin, N. F. and Abidin, S. Z. (2013), 'Upcycling: Reuse and recreate functional interior space using waste materials', *International Conference on Engineering and Product Design Education.*
Priya, A. (2021), 'Case study methodology of qualitative research: Key attributes and navigating the conundrums in its application', *Sociological Bulletin*, 70(1): 94–110.
Richardson, M. (2010), 'Design for reuse: Integrating upcycling into industrial design practice', *International Conference on Remanufacturing.*
Shuhailo, Y.V., Derkach, T.M., Shpetna. A.V. and Ustenko, T. (2022), 'Incorporation of upcycling techniques into technology education', *Journal Physics: Conference Series*, 2288(1): 012041.
Stahel, W. (2016), 'The circular economy', *Nature*, 531: 435–438.

Suwaed, H. (2022), 'Raising Libyan EFL undergraduate students' awareness about environmental issues through project-based learning', *Global Journal of Foreign Language Teaching*, 12(3): 112–124.

United Nations (2015), 'The 17 Goals – Department of Economic and Social Affairs', Available at: https://sdgs.un.org/goal.

Wu, J. and Bai, Z. (2022), 'Spatial and temporal changes of the ecological footprint of China's resource-based cities in the process of urbanisation', *Journal of Resources Policy*, 75: 102491,

Yin, R.K. (2014), *Case Study Research Design and Methods* (5th edn), Thousand Oaks: Sage Publications.

Zunaidah, A., Nathania, L. and Nurichsania, N.A. (2023), 'The Influence of green education on sustainable mindset in business English class', *3rd Asia Pacific International Conference on Industrial Engineering and Operations Management*.

Chapter 19

New Product Development for Upcycling and the Circular Economy Focusing on Everyday Electronic Products

Kyungeun Sung, Joseph Shade, Thomas Wylam, Tony Lorance, Mik Pieniazek and Abhishek Tiwary

Introduction

Many existing products on the market in the traditional linear economy (of take, make, use and dispose) have been designed and manufactured without much consideration of end-of-life options for the circular economy (turning waste into resources for production) (Stahel, 2016), such as remanufacturing or recycling (Burke et al., 2021). It is relatively easy and straightforward to reuse or recycle single-material products such as plastic bottles, metal cans, ceramic plates, or wooden furniture. Most everyday use electronic products such as electric kettles, speakers or vacuum cleaners, are made of mixed materials (e.g., several metals and plastics, rubber) and have multiple components that make them more difficult to remanufacture or recycle. Recognizing this challenge, this project aimed to create radical innovations in new product development for upcycling (utilizing used/waste materials for a new product with higher quality and/or value than the compositional elements) (Sung, 2017) and the circular economy focusing on everyday electronic products.

The project was a short-term project named 'New Product Development for Upcycling and Circular Economy' (NPD4CE), which ran between November 2022 and July 2023 and was funded by De Montfort University's (DMU) Higher Education Innovation Funding (HEIF) (£11,086). The project website is available at https://npd4ce.our.dmu.ac.uk/. We used the first two months (November-December 2022) to advertize the posts internally at DMU and externally via emails to design academics at Loughborough University and Nottingham Trent University. During this time, we also received applications, and shortlisted applicants. In the following month (January 2023), we interviewed the shortlisted students. We appointed three current students at DMU: Joseph Shade (BA Product Design), Thomas Wylam (BA Product Design) and Tony Lorance (MSc Product Design). During the administrative procedures (to hire students through DMU

Unitemps), we applied for and obtained research ethics approval for the project and had a project kick-off meeting involving the three students and the supervisory team: Dr Abhishek Tiwary (expert in sustainable engineering and manufacturing), Dr Kyungeun Sung (expert in upcycling and circular economy), and Mik Pieniazek (expert in product design practice and new product development).

The students were asked to choose a different product category. Joseph selected a speaker, Thomas selected a cordless vacuum cleaner, and Tony an electric kettle. For study purposes, we purchased the relevant and available benchmarking products for each category through DMU suppliers. From February to March 2023, students conducted literature reviews and other secondary research (e.g., benchmarking and identifying best practices), project planning, identification of relevant industry professionals/stakeholders and reverse engineering and product analysis of the purchased product. In April and May, the students generated ideas, developed concepts, interviewed the relevant industry professionals/stakeholders to capture industry perspectives and obtain feedback on their initial ideas and concepts, simulated and tested some of the promising concepts virtually and physically, and proposed a final concept. In July, they refined and detailed a final design proposal and created a working prototype. The actual time that the students spent on the project was a maximum of 10 hours a week (depending on other priority commitments such as assignment submission deadlines) for about 5.5 months between early February and mid-July of 2023. At the end of the project, we organized a project final presentation and networking event with the industry professionals in mid-July 2023 at DMU (Figure 20.1).

Figure 19.1 Presentation by Joseph Shade.

Student Project Processes and Outcomes

This section describes the three students' project processes and outcomes: Joseph Shade's speaker project, Thomas Wylam's cordless vacuum cleaner project, and Tony Lorance's electric kettle project. Because of the exploratory nature of the project, we allowed the students to develop their project in different directions.

Speaker Project by Joseph Shade

Joseph, in his secondary research, first tried to understand the key terms of upcycling and circular economy. Then, he described product life-cycle categories of product introduction to the market, growth, maturity and decline, and classified a speaker in the decline category. He then looked at design strategies for product design in the decline stage regarding sustainability: (i) design for product durability, (ii) design for standardization and compatibility, (iii) design for ease of maintenance and repair, (iv) design for disassembly and reassembly, (v) design for product attachment and trust, and (vi) design for upgradability and adaptability. In his primary research, he reverse-engineered and analysed the Sony SRS XB-23 Speaker (Figure 20.2) and identified several sustainability issues despite high-quality engineering and manufacturing overall. The issues included heavy use of glue, heavy soldering, a mixture of different materials, tiny screws requiring special drilling, and several snap hooks that were easily bent and broken. Clearly, the product was not designed for disassembly, repair or recycling.

Based on the secondary research and product analysis, Joseph had the initial idea of a classic, simple, longer-lasting design with customizable front-cover panels that are made of materials with graceful ageing (e.g., wood) and modular design in general for easy disassembly maintenance, repair and upgrade if required (Figure 20.3).

He talked to three product design practitioners with expertise in audio products, presented his initial idea and gained some feedback. Despite some contradictory opinions on details, the common opinion of the three professional designers was that the design should be timeless and desirable, promoting the upcycling narrative and targeting environmentally conscious consumers rather than audiophiles. Therefore, Joseph set his design brief as:

> The speaker will be easy to dis- and reassemble, pleasant to listen to yet not audiophile level and be highly modular. The target market will be the eco-conscious consumers. It will not use one-way fastening features, strongly avoid the use of uncommon fixings and be as durable as possible.

Focusing on strengthening the upcycling narrative, Joseph explored varied inspirations and decided to upcycle one meaningful waste (or found object) he possessed, a two-stroke engine barrel and waste wood. He first used 3D modelling and rendering to imagine the potential design outcome, then used 3D scanning of

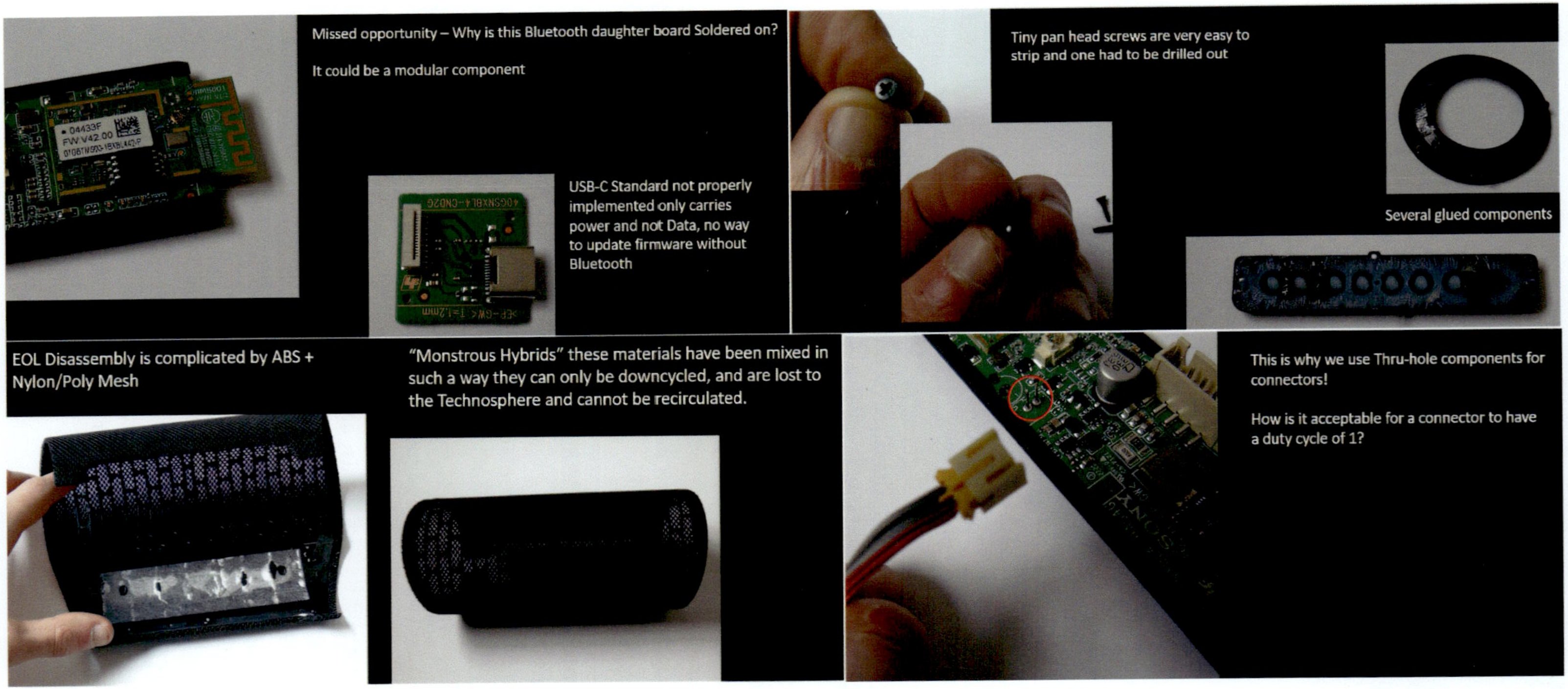

Figure 19.2 Sony SRS XB-23 Speaker reverse engineering and product analysis.

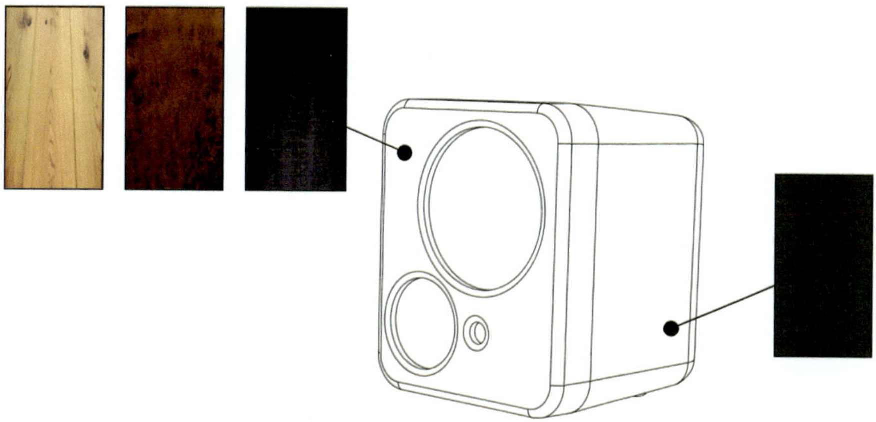

Figure 19.3 The initial idea of classic design with a customizable front panel.

the two-stroke engine barrel to get accurate measurements and build the design around it virtually and physically (Figures 20.4 and 20.5).

Joseph used two stereo speaker full-range drivers, BT-AMP, power supply, and speaker wire as working components, and a 3D-printed recycled plastic filament for the base (in addition to the two-stroke engine barrel and waste wood front panel) to create a working prototype (right photo in Figure 20.5). The original invention in his design is the new shape and configuration of the speaker using waste materials.

In his reflection, Joseph concluded that it is entirely possible to make a working product from reused components with few compromises. Despite the strong narrative of upcycling through particular aesthetics, which can be potentially more appealing to environmentally conscious consumers, due to the one-off nature of utilizing waste/found objects, the scalability of the concept in the commercial setting is questionable, requiring further research. Joseph's recommendation for similar future work was to (i) focus on narrative; (ii) stick to industry standards and conventions, especially for input/output, controls, and user-product interaction; (iii) design for long service life with repair and maintenance in mind; and (iv) consider the environment as another user group to consider their requirements.

Cordless Vacuum Cleaner Project by Thomas Wylam

Thomas, in his secondary research, first tried to understand the circular economy and other related concepts such as sustainability, planned obsolescence, and Life Cycle Analysis (LCA). Thomas focused on the UK's waste issue, including plastic waste export and undervalued reuse and recovery of materials. Then, he explained the significance of vacuum cleaners as a project's target product. For example, vacuum cleaners have the second largest embodied greenhouse gas emissions of

Figure 19.4 New upcycled speaker-making process.

Figure 19.5 New upcycled speaker design development.

electrical products after television and widespread ownership in households (Salvia et al., 2015). Vacuum cleaners are considered one of the fastest-growing waste streams and require specific enforcement legislation in the EU (Gallego-Schmid et al., 2016). Thomas introduced some good product design examples, such as a repairable flatpack toaster by Kasey Hou, which helps consumers to feel more attached to the product (so they use them for longer) and also feel confident in repairing it when it is broken by making the consumers assemble it at the start of ownership (Hou, 2023) – a widely identified phenomenon referred to as IKEA effect (Norton et al., 2012). Another good example was Kara, a modular coffee machine that is easy to repair and recycle, designed by Thomas Mair (Designboom, 2023). In his primary research, Thomas reverse-engineered and analysed Roidmi S2 – a cordless and bagless model (Figure 20.6). He found that whilst the assembly is

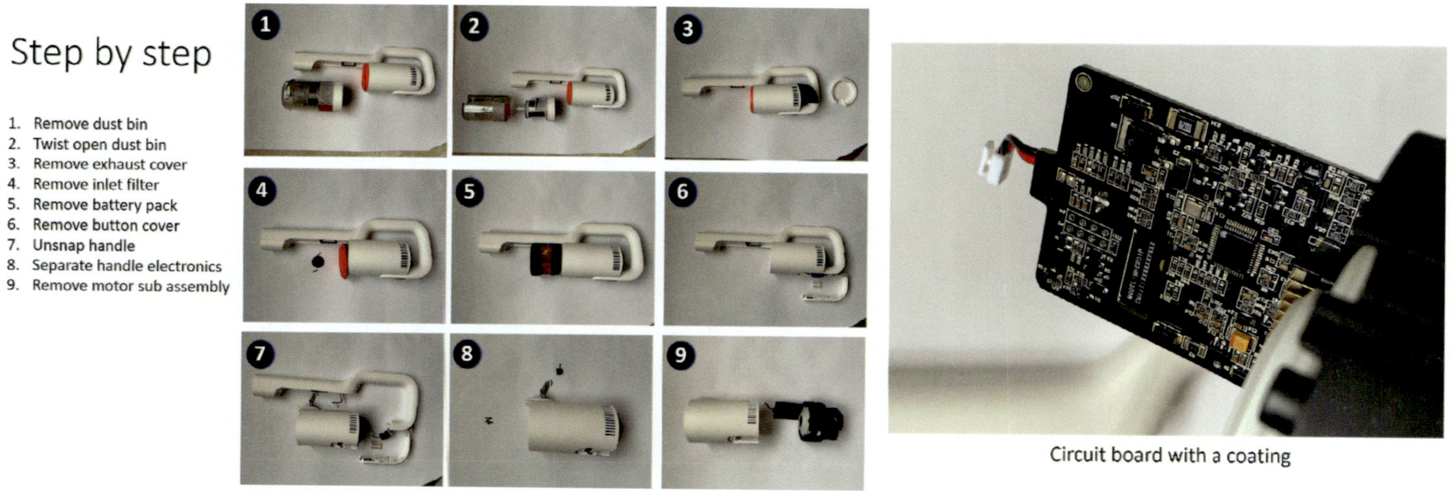

Figure 19.6 Roidmi S2 cordless vacuum cleaner reverse engineering and product analysis.

generally good with its use of a single screw type, there is a risk that some consumers could permanently damage the mouldings when undoing some snap hooks. The circuit board had a coating designed to prevent reverse engineering of the electronics or replacement of components during repair. The potential repair time was estimated as excessive (or longer than the industry average time) as the product had to be fully disassembled to access the motor and circuit board for a potential replacement.

From the research, Thomas identified several opportunities for design innovation that may help to extend the lifecycle and repairability of cordless vacuum cleaners, including designing appliances that maintain themselves (e.g., automated filter or brush bar cleaning) and designing to allow faster repair workflows. After experimenting with a variety of ideas, including some self-cleaning mechanisms, Thomas decided to go for a modular design for easy disassembly, repair, and reassembly, with protective features (i.e., detachable fins to absolve impact when dropped) for increased durability and potential upcycling (or creative reuse or repurpose) options after the service life such as a lampshade, plant pot, or toy (Figure 20.7). The originality of his design lies in the inclusion of modular protective fins.

Electric Kettle Project by Tony Lorance

Tony, in his research, lightly touched on the concept of upcycling and circular economy by literature review and focused on the selected/purchased product – Bosch TWK5P475GB 1.7L, a traditional electric kettle – for reverse engineering and product analysis (Figure 20.8). From the product analysis, Tony identified several sustainability issues: (i) use of press-fit components – once installed, they could not be removed without damaging the legs of the parts; (ii) non-removable heating coil, copper strips, and most of the electrical wirings – mostly welded to the container part; (iii) some fasteners requiring special tools for removal; and (iv) a complicated mechanism and high number of parts used for closing the lid.

Based on the identified issues, Tony proposed some design directions as follows:

- *Materials*: using durable and sustainable materials for the main body and handle, for example, recycled stainless steel, aluminium, bamboo or ceramic.
- *Accessibility*: making components more accessible for repair and maintenance by, for example, using standardized fasteners (not requiring special tools for removal).
- *Replaceability*: making the heating coil and copper strips more easily removable and replaceable (without welding) by, for instance, a plug-and-play mechanism.
- *Electrical wiring*: using removable connectors and detachable wiring.
- *Adhesives*: using sustainable adhesives such as non-toxic biodegradable ones.
- *Mechanism*: simplifying the mechanism for closing the lid and reducing the number of parts involved by, for example, using a magnetic or spring-loaded mechanism.
- *Fasteners and screws*: using those that are standard and easily accessible/removeable.

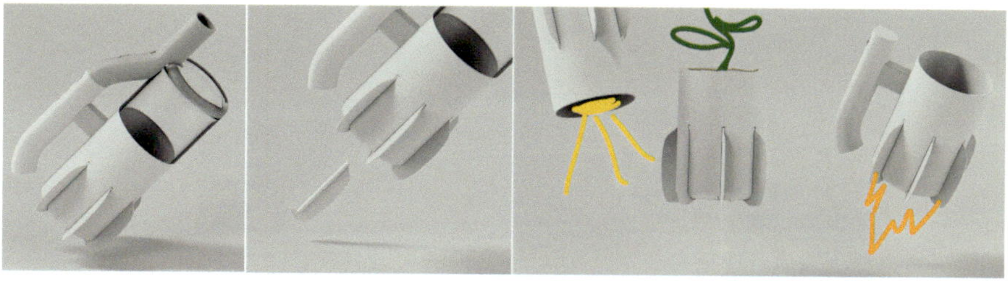

Figure 19.7 Modular, protective cordless vacuum cleaner for easy repairability and upcycling potential.

The design directions were in line with expert opinions. During the expert interviews, as Tony found out, they emphasized the need for standardizing fasteners and screws, redesigning the lid-closing mechanism and selecting more durable and sustainable materials. Tony, therefore, redesigned a kettle focusing on a simplified design with a more sustainable ceramic/glass body, standard fasteners and screws and wiring without welding. Tony created a working prototype (with the ceramic body) using some of the usable components from the reverse-engineered kettle, proving that it is possible to reuse all the components if it is designed and made for disassembly and reuse/remanufacture (Figure 20.9). Although the proposed design is visually indistinguishable from other existing kettle designs, the idea of improving the sustainability of the product by considering various aspects of the product is important to Tony's project.

Discussion and Conclusion

The three students who worked on the project found it to be a good learning opportunity. The industry professionals who were involved in the project throughout the process and who attended the final presentation and networking event also expressed how much they valued the opportunity to contribute to the project and be informed by new ideas. The results showed some potential for further research and development in this area. For example, we now know that speakers with an upcycling concept or a strong focus on sustainability are not for audiophiles and cannot easily be portable (if/when using various found/waste objects which are not necessarily lightweight), which limits the market to a great extent. Nowadays, not so many people use separate speakers at home unless they are audiophiles. Future research could investigate what kinds of products (or product categories) are more suitable for a strong upcycling concept or narrative. We know that furniture or fashion upcycling is popular to some extent (Singh et al., 2019), but what about other sectors or products? Where/what are the promising areas that have not yet been explored? On the other hand, the successful working prototype of the new speaker confirmed the technical feasibility of electronic products with upcycling concept. We just need to target the right market and

Figure 19.8 Bosch TWK5P475GB 1.7L electric kettle reverse engineering and product analysis.

Figure 19.9 Working prototype-making process and 3D modelling and rendering of the new kettle concept.

consumers. One of the critical feedback items from industry professionals regarding the new speaker design was the scalability of the concept based on a one-off design, which is a typical problem for many upcycled, handcrafted products. Unless we are targeting the high-end craft market, the manufacturing or production processes should be scalable (efficient enough to produce consistent quality results with controllable quantity) to make the business profitable. The next project will probably focus on scalable ideas and concepts that are ready for mass production.

Thomas's modular, protective cordless vacuum cleaner for easy repair and upcycling potential showed how current manufacturers could potentially create a significant impact by making a relatively small design change. The protective features (i.e., detachable fins), however, could be considered as non-essential components that may compromise the aesthetics and increase the manufacturing cost. Whether existing manufacturers would be interested in making these kinds of changes is unknown. If the newly suggested design change does not compromise the aesthetics (or any functions), if it does not add more parts/components and rather reduces the cost, manufacturers may be more willing to make that change. The suggested potential upcycling options (or creative reuse/repurpose) after the service life (such as a lampshade, plant pot, or toy) are certainly interesting; these are some of the most environmentally sustainable options as they do not use energy or toxic chemicals for processing. However, whether these options are the best ways to capture or increase the value of the existing products is debatable. Future studies could focus on making subtle changes that are not necessarily visible but could improve sustainability by, for example, reducing the number of parts, reducing the variety of materials and simplifying the assembly. The best practice identified from Thomas's research could have been another excellent design direction: a flat-pack, self-assembly cordless vacuum cleaner with the IKEA effect. Another future project could be looking at what kinds of products (or product categories) are more acceptable (to consumers or end users) for self-assembly.

Tony's kettle project showed that even a simple electric kettle has many small parts and components inside and the assembly is relatively complex. Therefore, there is still space for innovation regarding sustainability to improve the product design in terms of reducing and standardizing parts and simplifying the assembly.

The design direction Tony selected was promising – simplifying the design, using more sustainable materials, and making the disassembly and repair easier. However, the actual design and delivery were not as good as the ideas and concept. From the industry perspective, many manufacturers may have their business-as-usual practices using certain materials, fasteners or screws or welding certain parts. It could be challenging to change these current, prevalent practices as any change implies additional costs. To convince decision-makers of the manufacturers to make the change, we probably need to appeal for stronger financial benefits of the change or consider top-down interventions such as government policies or regulations or new industry standards emphasizing sustainability factors.

The project students were varied in terms of the stage of their study, capacity and strengths and weaknesses. Joseph was in his placement year after his second year in BA Product Design. As he was already in his part-time placement year to gain some work experience (without timetabled classes or coursework), he was the most passionate and enthusiastic person to dedicate his time and efforts to the project. In similar future projects involving student designers, the students who are in the part-time placement year might be the best people to recruit. Joseph's strengths were technical knowledge and capability, practical hands-on making and fabrication skills, 3D modelling and rendering and attention to detail. Thomas was in the final (third) year of BA Product Design. As he was working on his major project, degree show, and New Designers in London, he prioritized these more important study commitments, which was our recommendation, too. As an academic staff member, it is important to give a good learning opportunity to a variety of students, especially those who are going to be in the job market soon. However, if/when future design projects involving students require high level of commitment, one should avoid final-year students. Thomas's strengths were good storytelling and communication, a clear-thinking approach to problem-solving, a sophisticated taste in design and aesthetics and overall efficiency. Tony was an international student in MSc Product Design (unlike Joseph and Thomas, who were home students). As the master's programme is only one year, and someone in an unfamiliar environment (new to the UK and UK design education in higher education), Tony seemed to struggle a bit to balance between his study and this project work. So maybe it is not appropriate to involve an international master's student in any future projects unless it is one that is very short-term (e.g., one month). Tony's strengths were meticulous reverse engineering and product analysis, academic writing, an experimental approach to design, strategic and logical thinking, and conceptual design.

Based on these reflections, longer-term projects (e.g., 3–6 months) could be more suitable for second or placement year students. Shorter-term projects (e.g., 1–3 months) could also be suitable for final-year or master's students. Depending on students' strengths and weaknesses, it might be better to be flexible regarding the tasks so that each student has a tailored pathway or procedure to show their strengths and perform better in the project (instead of using the same procedures and methods). In our education, we emphasize flexible teaching and learning depending on students' diversity (Wade, 1994), and use it as an important principle in our teaching practice. However, when it comes to research projects, we sometimes forget that we could also apply flexibility to our projects so that the students who are involved in the project can learn and perform better.

The project ran for just over five months using our current students as part-time student designers with limited resources. The project was exploratory in nature and we do not claim that it has any scientific significance. However, we learned a lot from this project and this experience. We also hope that this chapter gives some useful tips to other academics who aim to conduct similar studies in the UK and elsewhere. This is a small-scale pilot project; we aim to scale this up by applying for larger amounts of external funding to address the questions raised by this project, including (i) products (product categories) suitable for strong upcycling concept; (ii) promising markets and consumer segments for products with strong upcycling concept; (iii) scalable upcycling design ideas and concepts suitable for mass production; (iv) subtle and invisible sustainable design innovation using upcycling principles; (v) products (product categories) suitable for self-assembly concept and IKEA effect; and (vi) how to involve and convince the big players (major manufacturers) for a change.

Acknowledgements

This work was supported mainly by De Montfort University's (DMU) Higher Education Innovation Funding (HEIF) and partially funded by the AHRC Research Networking grant for the International Upcycling Research Network [grant number AH/W007134/1]. This work was made possible by the generous Research Award from DMU. We thank Stuart Lawson, Kelley Wilder, Kate Cheyne, Siobhan Keenan, Christine White, Jill Cowley, Deborah Cartmell, Mike Kagioglou, and Katie Normington for supporting this research at DMU. Special thanks to the industry professionals who contributed to this project: Cameron Lalor, Charlie Keitch, Lewis Roe, Nathan Hales, and Seb Ward.

References

Burke, H., Zhang, A. and Wang, J.X. (2021), 'Integrating product design and supply chain management for a circular economy', *Production Planning & Control*, 34(11): 1097–1113.

designboom (2023), 'Repairable & recyclable Kara coffee machine sets a sustainable standard for appliance design', Available at: https://www.designboom.com/design/repairable-recyclable-kara-coffee-machine-sustainable-standard-appliance-design-thomas-mair-12-04-2022/.

Gallego-Schmid, A. Mendoza, J. M. F., Jeswani, H. K. and Azapagic, A. (2016), 'Life cycle environmental impacts of vacuum cleaners and the effects of European regulation', *Science of the Total Environment*, 559: 192–203.

Hou, K. (2023), 'Repairable flatpack toaster', Available at: https://www.kaseyhou.com/.

Norton, M.I., Mochon, D. And Ariely, D. (2012), 'The IKEA effect: When labor leads to love', *Journal of Consumer Psychology*, 22(3): 453–460.
Salvia, G., Cooper, T., Fisher, T., Harmer, L. and Barr, C. (2015), 'What is broken? Expected lifetime, perception of brokenness and attitude towards maintenance and repair', *Product Lifetimes and the Environment (PLATE) Conference proceedings*, 342–348, Nottingham: Nottingham Trent University.
Singh, J., Sung, K., Cooper, T., West, K. and Mont, O. (2019), 'Challenges and opportunities for scaling up upcycling businesses – The case of textile and wood upcycling businesses in the UK', *Resources, Conservation and Recycling*, 150: 104439.
Stahel, W.R. (2016), 'The circular economy', *Nature*, 531(7595): 435–438.
Sung, K. (2017), *Sustainable production and consumption by upcycling: Understanding and scaling up niche environmentally significant behaviour*, Nottingham: Nottingham Trent University.
Wade, W. (1994), *Flexible learning in Higher Education. Teaching and Learning in Higher Education Series*, London: Kogan Page Limited.

Chapter 20

Common Waste Upcycling for Product Design as Extracurricular Activities in Design Education

Kyungeun Sung, Millie Rees, Theodore Sloos, Richie Moalosi, Yaone Rapitsenyane and Anneloes Roelandschap

Introduction

Upcycling of common wastes such as plastics, textiles and wood is known to be a promising area in terms of reducing negative environmental impacts and becoming a financially viable option as an alternative production method (Sohn et al., 2020; Singh et al., 2019). Common waste upcycling for product design or new product development is highly relevant and important in line with the United Nation's Sustainable Development Goal 12, Responsible Consumption and Production and the emerging circular economy – an alternative to the traditional, linear economy of take, make, use and dispose, emphasizing material circularity (Stahel, 2016). Embedding upcycling, circular economy, and responsible consumption and production in design education is one of the first steps to making changes in industry practices (Sung, 2023). When it comes to upcycling for teaching and learning in higher education, a recent literature review (Sung, 2020) found that most articles were published between 2014 and 2018 by academics in the fields of business management, design, sustainability science and engineering, confirming the nascent stage of research in this area. As part of our contribution to the body of literature, as well as making a difference in education and, ultimately, industry, we used De Montfort University's internal student internship programme (the annual Frontrunner Internship) to carry out two upcycling projects between 2021 and 2023: one project between October 2021 and March 2022, and another between October 2022 and Mach 2023. Two BA Product Design students from De Montfort University were employed on a part-time basis for the two projects: Millie Rees for 2021–22 and Theo Sloos for 2022–23.

Frontrunner Internship with Upcycling

De Montfort University (DMU) offers the annual Frontrunners internship scheme which gives current students the opportunity to undertake challenging employment and develop the higher-level skills necessary to compete for the best graduate jobs. Internships are normally 21 weeks in length and 10 hours per week, taking place during term-time only. The expectation is that internships offer a secure and supportive work environment with on-the-job training and a high level of supervision and mentoring. Internships are developmental, with the main emphasis on the benefits to the employee. The first author (as a Senior Lecturer in Product Design at DMU) applied for and secured two internship positions for the academic year 2021–22 and 2022–23 (internships titled 'Computer Aided Design and Upcycling'). The internship opportunities aimed to target students who would like to pursue a design career using computer-aided design and/or employing sustainable design principles closely related to upcycling – the creative process of turning wastes into high-quality/value products (Sung, 2017). The main duties of the internships varied, yet the one related to upcycling was the 'design and creation of new upcycled product(s) for commercialisation'.

Project Processes and Outcomes

Millie's project in 2021–22 focused on textile upcycling; she worked independently and autonomously throughout the project. On the other hand, Theo's project in 2022–23 had glass, plastic and wood upcycling and he worked with varied external partners from the University of Botswana (Industrial Design and Technology) and Chako Zanzibar – a sustainable product design and production social enterprise-based on upcycling (https://chakozanzibar.com/). The following sub-sections explain each project's processes and key outcomes.

Textile Upcycling Project by Millie

The first upcycling task in the project for Millie was to produce some upcycled products for the DMU Made Christmas Fair 2021. Millie used old bedding to upcycle into the cushions and door sausages. She used cardboard cut-out templates from waste pizza boxes to make the template for the cushions. She also made stuffed animals using textile waste. Several textile upcycling products were created and made available for the DMU Made Christmas Fair 2021 (Figure 21.1). At the fair, only one stuffed animal and one door sausage were sold. However, since the publicity from the fair, Millie made three more sales and had two more orders through her online account for cushions in December 2021. She also had a few requests for cushions in the shape of letters as presents for people for Christmas. Millie created B cushions for the lady who wanted a present for her son named Ben (Figure 21.2). In her reflective journal, Millie stated:

> If I did the DMU Christmas fair again, I would look at expanding the variety of my products. I really enjoyed the fair, but I did not sell very much, and the products that I did sell were more unique than the other things available at the fair. If I did it again, I would do more experimentation [...] and look at adding

Figure 20.1 DMU Made Christmas Fair.

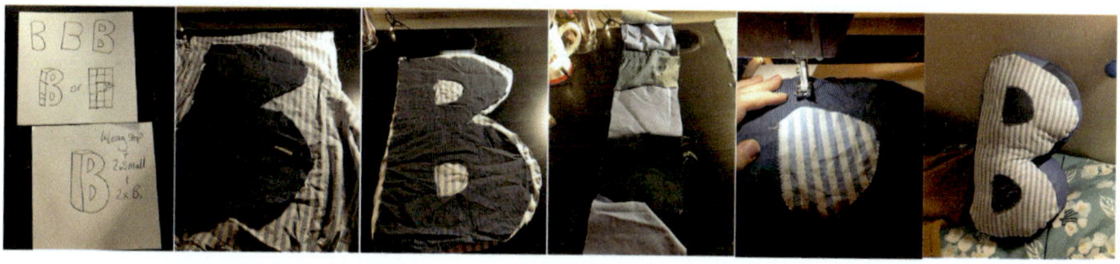

Figure 20.2 Personalized/customized letter upcycled cushion.

> more personalised branding to the products like a label to show that they have been recycled from something else.

The second upcycling task for Millie was to organize a hands-on upcycling workshop for the public in Leicester city centre in February 2022 as part of the launch event of the Beta X – Leicester-based design and sustainability hub for a pop-up programme of digital work, fashion, textiles, film, photography, carnival, makers, visual arts, demonstrations, workshops, talks, events and more. She ran two two-hour workshops on 19 February 2022. Each workshop had about 15 local community people as participants. Millie provided the participants with cardboard cut-out templates (new pattern cutting made by Mille), waste textiles, sewing machines, stuffing materials and other relevant tools to create stuffed whales. She taught the participants how to operate the sewing machines. All the participants seemed to be highly engaged and managed to create their own unique textile pattern whales (Figure 21.3). In her reflective journal, Mille mentioned:

> Although it was a little stressful some of the time, I found it really gratifying watching my designs go into production, being created by other people and seeing the same excitement on their faces from the same product that I had first designed at midnight in my bedroom! I learned the value of accepting help from other people during the workshop. If I were to run the workshop again in the future, I would bring people to help me out with the workshop as it was difficult

to control what was going on when multiple people got to a difficult stage of making the whale. […] Since the workshop, I have been approached [by someone] to do another upcycling or pyjamas workshop […] Even though I could not do this workshop [due to the time limit], it was still great feedback on the workshop to be asked back to put on another one. At some point in the future, I would love to put on more workshops and to spend time teaching other people how to design and upcycle animals and products.

Experimental Upcycling Project with Varied Materials by Theo

One of Theo's upcycling tasks was to create high-quality furniture out of pallets with design and manufacturing processes that can be repeated in Botswana, targeting street vendors in Gaborone (the capital city). The initial idea came from the design academics working on another pallet upcycling project with students at the University of Botswana. Theo obtained a spare pallet from the industrial estate near De Montfort University. As the pallet was wet and dirty, Theo cleaned and dried it first. He generated some furniture ideas, including Russian stools, lounge chair, and folding garden chairs. During the first feedback meeting with the academics from the University of Botswana, the design direction was decided to be the lounge chair. Theo sanded and cut the pallets into pieces to build the structure/frame of the chair. Once the structure was created, he worked with a textile technical tutor to make tailored back and seat pads for the chair. He sprayed

Figure 20.3 Upcycling workshop as part of Leicester Beta X.

Figure 20.4 Pallet upcycling chair design and creation process.

the chair frame (sky blue) and dyed the pads (black) to match the colour theme of the flag of Botswana (Figure 21.4). Theo also made an IKEA-style assembly/building instruction document for potential future reproduction (Figure 21.5).

One of the original plans was for Theo to deliver a short online guest lecture to the Industrial Design and Technology students at the University of Botswana if he created exceptionally high-quality reproducible furniture. However, the quality was deemed unsatisfactory by the client (the University of Botswana). Therefore, we dropped the idea. Despite the disappointing outcome from the perspective of the client, the student seemed to value this special learning opportunity, which is mostly evidenced in his reflective journal:

> This was a brilliant opportunity for design work with a client, creating experience for consultancy-like work. I learned a great deal from this, though I felt there were some limitations. The main difficulty was communication. […] Another brilliant skill that was developed was trying to create understandable instructions for a layman to follow. I produced IKEA-like instructions, which helped me better understand my own process of design, as well as consider how the product could be made simpler for easier assembly.

Theo also worked with Chako Zanzibar in Tanzania – a social enterprise based on upcycling design and making (https://chakozanzibar.com/). Theo was asked to assist with creating new ideas related to Chako's existing set of products. He experimented with different materials (e.g., cardboard boxes, glass bottles and plastic) for different purposes. Although none were usable for commercial purposes, they showed some interesting potential for future development, such as the weaving of different waste materials to create something. In his reflective journal, Theo stated:

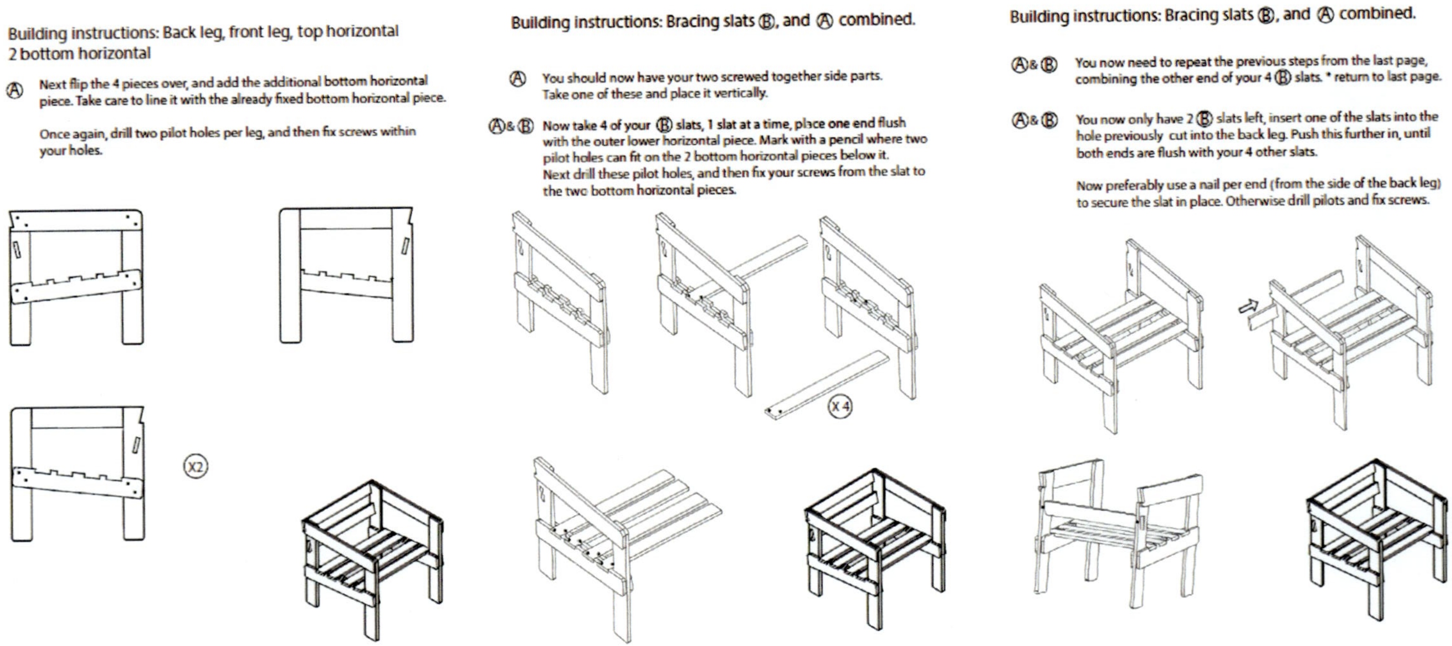

Figure 20.5 IKEA-style building/assembly instruction.

> I was asked to assist with their new idea of producing hotel wristbands out of extruded plastic. This led me to learn how to sandblast glass, creating press cuts. The other project that was brought forward was to create a plastic woven lampshade. This was greatly out of my comfort zone with weaving, and I struggled to get the project off the ground. I made multiple prototypes but became frustrated and struggled to continue pursuing the ideas. [...] I felt the project would have been more adequately suited to someone with more artistic and soft crafting skills. I am glad I had the experience as it has helped me to realize that weaving isn't something that suits me.

Discussion and Conclusion

The outcomes of the two projects were mainly relevant to and significant for the student's learning experience rather than creating a big commercial success or showcasing best practice. Therefore, these two projects achieved their original primary aim. The projects were meaningful extracurricular activities in design education in the higher education context not only to enhance the students' employability but also to impart additional knowledge and skills related to upcycling and the circular economy to future designers who could potentially transform (or at least influence) our future industries. We also have learned lessons and gained insights from the projects. For example, from the rather unsuccessful Christmas fair and Millie's reflection on it, similar future endeavours could aim to generate and produce a variety of unique upcycling products with more experimental approaches to design and creation. It would be better to have many one-off unique products if they are based on manual crafting and making and aimed at the Christmas fair. Branding and labels to indicate the original waste materials and the degree of sustainability may help the sales. Personalization or customization options could be provided to attract certain customers. Based on the relative success of public engagement in upcycling workshops, we can safely recommend that other academics engage students in creative art and design programmes for similar events. Despite the student's satisfactory experience with the upcycling workshops (based on great engagement and positive feedback), it was still a stressful event due to the heavy workload. Similar hands-on practical upcycling workshops in the future anticipating 10 to 15 participants should have at least two people to assist the participants.

From the rather disappointing outcome of the pallet upcycling furniture project, we could see the limitations of the limited capacity and capability of the designer/maker. Similar future projects (targeting furniture or craft-based upcycling) should hire students who have a higher level of hands-on making and fabrication skills. It sounds obvious, but it was something we did not see at the beginning of the project, given our assumption that product design students have sufficient workshops and practical hands-on skills in general. Overall, however, we are glad that the student had a great learning opportunity. Inspired by Theo's trial and error, we believe that there is some potential for future projects such as pallet upcycling for smaller-scale products with a high-quality finish which can be commercialized (e.g., toys, home decorations), a range of sustainable flat-pack products (outside furniture) with self-building/assembly instructions and weaving a number of flexible waste material products together to make a product (e.g., coasters).

After these two projects, we (Product Design at DMU) decided to use our students' strengths for future Frontrunners Internship schemes (or any other extracurricular activities). Therefore, in future, we will probably focus on new product development or redesigning existing products as incremental innovation aiming for upcycle-ability or the circular economy, moving away from craft-based upcycling. We also strongly encourage other academics to use their own students' strengths as the starting point to form or shape any extracurricular activities and to supervise and advise students based on their strengths. It would still be good to involve external partners as clients in the projects, preferably more local clients, so that it is relatively easy to get regular feedback on the processes and outcomes, and communication can be more efficient and effective. The best approach would be to make this a part of the CSR (Corporate Social Responsibility) activities of local companies/clients for improved sustainability for their existing products or services.

We academics usually report on success stories (or successful studies) but are often hesitant to report on unsuccessful cases, doubting their intellectual value. In fact, people can learn from both their successes and their failures (Ellis et al., 2014). We also learn from other people's successes and failures. We believe that this chapter contains both reasonable successes and good failures from which there is much to learn and illustrates many useful do's and don'ts. We hope that this chapter informs and inspires other academics in design education in higher education (or other educational settings) in their future planning for extracurricular activities focusing on sustainability.

Acknowledgements

This work was supported mainly by the De Montfort University (DMU) Frontrunners Internship scheme and was partially funded by the AHRC Research Networking grant for the International Upcycling Research Network [grant number AH/W007134/1]. This work was possible with the generous Research Award from DMU. We thank Nicky Harding, Stuart Lawson, Kelley Wilder, Kate Cheyne, Siobhan Keenan, Christine White, Jill Cowley, Mike Kagioglou, and Katie Normington for supporting this research at DMU. Special thanks to James Burkmar at LCB (Leicester Creative Business) Depot for supporting the upcycling workshops in Leicester Beta X and the Leicester community people who participated in the workshops to make the event a success. We also thank our workshop technicians, technical tutors and other Product Design academics at DMU for helping Theo.

References

Ellis, S., Carette, B., Anseel, F. and Lievens, F. (2014), 'Systematic reflection: Implications for learning from failures and successes', *Current Directions in Psychological Science*, 23(1): 67–72.

Singh, J., Sung, K., Cooper, T., West, K. and Mont, O. (2019), 'Challenges and opportunities for scaling up upcycling businesses – The case of textile and wood upcycling businesses in the UK', *Resources, Conservation and Recycling*, 150: 104439.

Sohn, Y.J., Kim, H.T., Baritugo, K.A., Jo, S.Y., Song, H.M., Park, S.Y., Park, S.K., Pyo, J., Cha, H.G., Kim, H., Na, J.G., Park, C., Choi, J.I. and Joo, J.C. (2020), 'Recent advances in sustainable plastic upcycling and biopolymers', *Biotechnology Journal*, 15(6): 1900489.

Stahel, W.R. (2016), 'The circular economy', *Nature*, 531(7595): 435–438.

Sung, K. (2017), *Sustainable production and consumption by upcycling: Understanding and scaling up niche environmentally significant behaviour*, Nottingham: Nottingham Trent University.

Sung, K. (2020), 'Upcycling for teaching and learning in higher education: Literature review', in W.L. Filho, A.L. Salvia, R.W. Pretorius, L.L. Brandli, E. Manolas, F. Alves, U. Azeiteiro, J. Rogers, C. Shiel, A.D. Paco (eds), *Universities as living labs for sustainable development: supporting the implementation of the sustainable development goals*, 371–382, Cham: Springer.

Sung, K. (2023), 'Understanding upcycling and circular economy and their interrelationships through literature review for design education', *Proceedings of the Design Society*, 3721–3730.

Chapter 21

Integrating Upcycling in the Industrial Design Programme as an Alternative Approach to Mitigate Environmental Challenges in Botswana

Thatayaone Mosepedi and Gerald Kennedy

Introduction

The growing concern for environmental sustainability has led to a renewed interest in alternative approaches to mitigate environmental challenges, particularly in developing countries. The suggested approach involves integrating upcycling into the industrial design programme to promote sustainable practices, stimulate innovation and reduce waste generation. Upcycling refers to the creative process of transforming waste materials or discarded products into new materials or products of higher quality, functionality, or environmental value. This concept has gained popularity in recent years due to the growing concern for environmental sustainability and the urgent need to mitigate the negative impacts of waste on the planet (Ghisellini et al., 2016). Despite the numerous benefits of upcycling, it has not received much attention from institutions of higher learning in Africa, particularly in the fields of design and manufacturing (Bennett, 2019). This chapter aims to investigate the potential of integrating upcycling into the industrial design programme in Botswana. This country faces numerous environmental challenges, such as waste management, deforestation and water scarcity (DEA, 2015).

The essence of this exploration lies in adopting a mixed-methods approach that seamlessly interweaves quantitative surveys and qualitative interviews to tap into the insights of educators, students, and industry professionals. By unearthing the multifaceted dimensions of integrating upcycling into industrial design education, this study aspires to illuminate the potential avenues through which environmental concerns can be addressed, innovative examples can be ignited and sustainable practices can be disseminated within the distinctive contours of Botswana's environment. It is important that this research not only highlights the potential of upcycling but also identifies and addresses the challenges that may prevent its seamless integration into the curriculum. Key challenges include lack of awareness, inadequate infrastructure and financial constraints. Thus, the subsequent sections of this chapter delve into an exploration of the various dimensions, challenges, and recommendations concerning the integration of upcycling in the industrial design programme. The ensuing discussion explains the pivotal role of upcycling as an alternative approach to bolstering the nation's sustainable trajectory, opening a new chapter in the intersection of industrial design and ecological stewardship.

Problem Statement

The concept of upcycling has gained considerable attention in recent years, as it offers a promising solution to waste management and environmental challenges. By turning waste materials into new, valuable products, upcycling can reduce the demand for raw materials, decrease greenhouse gas emissions and create job opportunities. Several studies have highlighted the benefits of integrating upcycling into industrial design education, including enhanced creativity, innovation and problem-solving skills (Baldwin et al., 2018; Chinda & Suri, 2020).

However, within the realm of industrial design education, the incorporation of upcycling remains an underexplored avenue. The lack of attention to upcycling in design curricula provides students with an incomplete narrative, one that avoids the pivotal role that upcycling could play in fostering sustainable practices and nurturing a culture of conscious design. As the tide of environmental consciousness sweeps across the global landscape, industrial design programmes have a profound opportunity to shape the next generation of designers as stewards of sustainability. This chapter aims to bridge this gap by investigating the potential of integrating upcycling into the industrial design programme in Botswana, thereby contributing to a richer understanding of the challenges, prospects and pathways for this transformative integration.

The environment of Botswana presents an intricate tapestry of challenges that necessitate innovative and adaptive solutions. The country grapples with the complicated interplay of waste management, deforestation and water scarcity (DEA, 2015). Against this backdrop, the exploration of upcycling becomes imperative. By delving into the feasibility of integrating upcycling within the industrial design curriculum, this study responds to the unique demands and aspirations of Botswana's environmental narrative. The alignment of design innovation and ecological consciousness holds the potential to sculpt a sustainable industrial landscape that aligns with Botswana's developmental goals.

Purpose of the Study

The research seeks to evaluate the effectiveness of incorporating upcycling principles in the Department of Industrial Design and Technology and assess the impact on students' design projects. The purpose of this study is to explore the potential benefits of integrating upcycling into the industrial design programme in institutions of higher learning in Botswana (University of Botswana) and analyse the impact it could have on addressing environmental challenges in the country.

Literature Review

While the concept of upcycling has gained prominence in recent years as a promising approach to addressing environmental challenges, a closer examination of the literature reveals both its potential benefits and the gaps in its implementation within the field of industrial design education. Upcycling is also closely related to the concept of sustainability, which seeks to balance the needs of the present without compromising the ability of future generations to meet their needs (Brundtland, 1987). The practice of upcycling contributes to sustainability by minimizing waste, conserving resources, and promoting a circular economy (Andersen, 2007). The circular economy is an economic system that aims to eliminate waste and promote the continuous use of resources through closed-loop systems, ultimately reducing the environmental impact of production and consumption (Geissdoerfer et al., 2017). This transformation has implications beyond material reuse; it extends to enhancing resource efficiency, promoting creative problem-solving, and fostering a shift towards more responsible production and consumption patterns.

A significant number of products that have reached the end of their lifetime in Africa become waste, which may lead to environmental degradation (Hoornweg & Bhada-Tata, 2012). Although the benefits of upcycling in product design are evident, there is very little research to validate it (Bennett, 2019). Meanwhile, the available research in the area is aligned with the textile and fashion industry (Fletcher, 2014). Upcycling can help minimize the products' effects on the planet, ranging from reducing strain on valuable resources such as minerals and fossil fuels to reducing the hazardous waste produced when new products are manufactured (Ghisellini et al., 2016). Its implementation in industrial design education faces several challenges, such as lack of awareness, inadequate infrastructure and limited funding (Chinda & Suri, 2020). Moreover, there is a scarcity of research on the integration of upcycling in the industrial design programme in developing countries, particularly in Africa.

In the context of Botswana, the industrial design programme is still in its infancy; there is a pressing need to incorporate sustainable practices into the curriculum to address the country's environmental challenges. Upcycling, as an alternative approach, can contribute to the development of a sustainable, eco-friendly industry in Botswana by promoting circular economy principles, such as waste prevention, resource efficiency and closed-loop production.

The literature review underscores the potential of upcycling as a sustainable approach within industrial design education. While existing research highlights the environmental benefits and potential innovation that upcycling can bring, the scarcity of studies focusing on its implementation in African contexts, particularly in industrial design education, points to an unexplored research niche. This study aims to bridge this gap by investigating the potential of integrating upcycling practices into the industrial design programme in Botswana, contributing to both the body of knowledge and the practical application of upcycling principles within a unique context.

Research Method

The study followed an exploratory sequential mixed-methods approach, combining both qualitative and quantitative data to ensure a comprehensive understanding of the research problem (Creswell & Creswell, 2018). The sequential engagement with qualitative interviews and quantitative surveys ensures that both the depth and breadth of insights are unearthed. This methodological synergy allows for a comprehensive examination of the potential benefits, challenges and prospects associated with the integration of upcycling into the industrial design programme.

Participants and Sampling

The participants for the qualitative phase (Phase 1) of the study were selected using purposive sampling, which involved selecting participants based on their expertise, experience or knowledge of upcycling and industrial design (Palinkas et al., 2015). The target population included industrial design educators, students and practitioners, as well as representatives from environmental organizations and relevant government departments in Botswana. A total of 15 participants were interviewed, ensuring diversity in perspectives and experiences, culminating in a nuanced understanding of the potential of upcycling within industrial design education.

The target population for the quantitative phase (Phase 2) of the study were students enrolled in industrial design programmes in institutions of higher learning in Botswana. A stratified random sampling technique was employed to ensure the sample was representative of the target population. The strata were determined by the level of study (undergraduate level 300 and 200, respectively). This stratification ensures that the sample is representative of the student body, offering an holistic view of attitudes, knowledge and perceptions across different stages of education. A total of 41 students were selected to participate in the survey, with a response rate of 68.3 per cent (n=28).

Phase 1 Qualitative Data Collection and Analysis

Semi-structured interviews and focus group discussions were employed as data collection methods for the qualitative phase. These methods were chosen for their capacity to delve into participants' perspectives, experiences and insights, providing a rich tapestry of qualitative data. The semi-structured nature of the interviews and discussions allowed for flexibility, enabling participants to elaborate on their viewpoints and share unique insights. Each interview and focus-group discussion was conducted using a pre-determined interview guide, which contained open-ended questions related to the research objectives (Creswell & Creswell, 2018) that invited participants to reflect on their perceptions of upcycling, its potential benefits, challenges and the feasibility of integrating it into the industrial design curriculum. This ensures consistency and comparability across interviews and discussions. Throughout the data collection process, the emphasis was placed on establishing rapport with participants and fostering an open and candid dialogue. Ethical considerations were paramount, with participants' consent obtained before the initiation of data collection. The interviews were audio-recorded to capture participants' responses accurately and transcribed verbatim to ensure accurate data representation to facilitate the subsequent analysis.

Thematic analysis was employed to analyse the qualitative data collected from the interviews and focus group discussions. This approach involved identifying and categorizing patterns or themes within the data (Braun & Clarke, 2006). The analysis followed a structured yet iterative approach, beginning with familiarization of the data, followed by generating initial codes, identification of emerging themes, reviewing themes, defining and naming themes, and producing a final report. The thematic analysis yielded insights into participants' perceptions of upcycling, its potential advantages, challenges and the strategies deemed necessary for its effective integration into the industrial design curriculum.

Phase 2 Quantitative Data Collection and Analysis

A self-administered questionnaire survey was used to collect quantitative data. The questionnaire consisted of four sections: (i) demographic information, (ii) knowledge and awareness of upcycling, (iii) attitudes towards upcycling in industrial design, and (iv) perceived barriers and facilitators to integrating upcycling in the curriculum. The questionnaire was distributed online using a secure digital platform and was completed anonymously by the participants. Throughout the quantitative data collection phase, careful attention was given to the clarity of the questions, the representativeness of the sample and the ethical considerations of data collection from all participants.

Descriptive statistics, such as frequencies, percentages, means, and standard deviations, were calculated to summarize the demographic information and responses to the questionnaire. Analysis of variance (ANOVA) was used to investigate relationships between variables and to compare means between groups (Field, 2013). All statistical analyses were conducted using the Statistical Package for the Social Sciences (SPSS) software version 28, which serves as the analytical engine, enabling systematic and rigorous analyses of the quantitative data. The data analysis process adheres to robust statistical principles, ensuring the reliability and validity of the interpretations drawn from the dataset.

Findings

Qualitative Findings

The thematic analysis of the qualitative data revealed four major themes: (i) the benefits of integrating upcycling in the industrial design programme, (ii) challenges in implementing upcycling, (iii) strategies for successful integration of upcycling, and (iv) the role of stakeholders in promoting upcycling in industrial design education. The first theme, 'benefits of integrating upcycling in the industrial design programme', underscored the perceived advantages of incorporating upcycling principles into industrial design education. Participants identified a range of benefits, including enhanced creativity, innovative problem-solving skills and a heightened understanding of sustainable practices. The capacity of upcycling to instil a sense of purpose and environmental stewardship in students was a recurrent theme. The second theme, 'challenges in implementing upcycling', uncovered the potential barriers that cast shadows over the seamless integration of upcycling into the curriculum. Lack of awareness, limited resources and inadequate infrastructure emerged as significant impediments. Participants also highlighted challenges related to the assessment and evaluation of upcycling projects within the existing educational framework.

The third theme, 'strategies for successful integration of upcycling', delves into the pragmatic approaches that can foster the effective integration of upcycling into the industrial design programme. Capacity building for educators, interdisciplinary collaborations, and the establishment of dedicated workshops were advocated as strategies to navigate the challenges. Participants also underscored the importance of community engagement and the involvement of external stakeholders. The fourth theme, 'the role of stakeholders in promoting upcycling in industrial design education', elucidates the pivotal roles that different actors play in promoting upcycling within the context of industrial design education. Government agencies, environmental organizations, industry partners and academia were identified as critical stakeholders that could facilitate the integration process. The significance of collaborative efforts and cross-sector partnerships emerged as a key insight.

Quantitative Results

The survey results indicated that most of the students (91.9%) were aware of upcycling and its potential benefits. However, there was a significant knowledge gap in terms of practical implementation and integration in the curriculum. Positive attitudes towards upcycling in industrial design were reported (83.9%), with students expressing interest in learning more about upcycling techniques and practices (87.3%). Perceived barriers included lack of resources (22%), limited knowledge among educators (41%), and inadequate infrastructure (37%). Conversely, facilitators encompassed training and capacity building for educators (53%), increased awareness campaigns (38%), and support from relevant stakeholders (32%). These quantitative findings align with the qualitative insights, reinforcing the need for targeted interventions to overcome challenges and leverage opportunities.

Discussion

The findings of this study resonate at the intersection of sustainability, design education and the industrial landscape. The integration of upcycling principles into the industrial design programme holds the promise of catalysing multifaceted transformations, addressing environmental challenges, and fostering a culture of conscious creation. Upcycling can address the following:

- *Environmental challenges*: The potential of upcycling to address Botswana's pressing environmental challenges cannot be overstated. The nation's struggle with waste management, deforestation, and water scarcity necessitates innovative approaches that harmonize human activities with the natural world. The integration of upcycling emerges as a beacon of hope, offering a route towards waste reduction, resource conservation and the development of eco-innovations that alleviate environmental stressors.
- *Fostering sustainable design practices*: The symbiotic relationship between upcycling and sustainability is illuminated through the lens of industrial design. The marriage of design innovation and ecological consciousness ushers in a new era of design thinking – one that transcends mere aesthetics to embrace the principles of circularity, resource efficiency and waste reduction. The positive attitudes exhibited by students towards upcycling underscore the potential for this integration to not only shape their design projects but also cultivate a lasting commitment to sustainable design practices.
- *Overcoming challenges through collaborative efforts*: The challenges identified in this study, including lack of awareness, limited resources and inadequate infrastructure, resonate with broader narratives surrounding sustainable education and practice. However, these challenges are not insurmountable. The study's insights on strategies for successful integration, such as capacity building, interdisciplinary collaboration, and stakeholder engagement, offer actionable pathways to navigate these hurdles. The role of stakeholders – government agencies, industry partners, academia, and environmental organizations – takes centre stage as catalysts for change.
- *Pathways to a transformed industrial landscape*: The integration of upcycling principles not only resonates within the confines of educational institutions but extends its tendrils into the broader industrial landscape. The study's findings illuminate the role of government policies and regulations in nurturing a conducive ecosystem for upcycling. Supportive policies could pave the way for a transformed industrial sector that champions sustainable design, thereby redefining the contours of Botswana's economic and ecological narrative.

Recommendations

The insights gleaned from this study culminate in a series of recommendations that collectively form a blueprint for harnessing the transformative potential of upcycling within the industrial design programme. These recommendations serve as guideposts for institutions, educators, policymakers, industry partners and students alike as they embark on a journey of sustainable innovation and conscious creation. Based on the study findings, the following recommendations are proposed.

- *Curriculum development and revision*: Institutions offering industrial design programmes should integrate upcycling concepts and practices across various courses, exposing students to the philosophy and applications of upcycling. This integration nurtures sustainable design mindsets.
- *Educator capacity building*: Conduct workshops and training sessions to equip educators with upcycling knowledge and pedagogical approaches, amplifying the impact of upcycling integration.
- *Awareness campaigns and advocacy*: Through workshops, raise awareness about the benefits of upcycling and encourage students to champion upcycling principles within and beyond academia.
- *Collaborative partnerships*: Forge partnerships with government agencies, industries, and community stakeholders to support upcycling workshops and student-led projects.
- *Resource allocation and infrastructure*: Allocate resources for workshops equipped with tools and materials for upcycling projects, highlighting the institution's commitment to sustainable design practices.
- *Research and exchange programmes*: Establish international research and exchange programs to expose students to diverse upcycling practices and cross-cultural perspectives.
- *Continuous assessment and evaluation*: Establish mechanisms for tracking the effectiveness of upcycling initiatives, adapting strategies based on student feedback and industry needs.
- *Platforms for knowledge sharing*: Create online repositories and events dedicated to upcycling, fostering innovation, and sharing best practices.
- *Advocacy for supportive policies*: Advocate for policies promoting upcycling and sustainable design practices within the industrial sector.
- *Ongoing reflection and adaptation*: Encourage continuous improvement, flexibility, and responsiveness to emerging challenges in sustainable design education.

Limitations and Future Research

While this study provides valuable insights into the integration of upcycling within the industrial design programme, there are several limitations that need to be considered. These limitations could pave the way for future research that can deepen our understanding of the impact of upcycling on sustainable design education. The study's narrow focus on one location and institution limits how broadly we can apply its findings. Future research should encompass diverse institutions and countries to offer a more comprehensive view. While the study looked at students and educators, other stakeholders, such as policymakers, industry partners, and community members, have crucial insights. Future research should include these voices for a more holistic understanding. There is a need to explore the lasting effects of upcycling education on students' practices and mindsets. Longitudinal studies could reveal how it influences careers, sustainable design, and graduates' contributions to sustainability. Going deeper into teaching methods and learning outcomes would clarify how best to integrate upcycling into design education. Analysing strategies across institutions and countries would help identify best practices and challenges, fostering cross-cultural collaborations in sustainable design education. Understanding industry expectations and perceptions of design graduates with upcycling expertise could inform collaborations between academia and industry. Emphasizing quantitative analyses could validate qualitative insights and offer statistical generalisability to findings. Exploring the ecological footprint of upcycled design projects could illustrate the tangible contributions of design graduates to environmental sustainability within the industrial design programme.

Conclusion

The integration of upcycling into the industrial design programme within the context of Botswana's unique environmental challenges offers a promising pathway towards sustainable development and innovation. This study's findings underscore the multifaceted potential of upcycling to transcend the boundaries of traditional design education and catalyse transformative change within the industrial landscape. The infusion of upcycling principles into the curriculum resonates as a transformative force capable of addressing the pressing environmental challenges faced by Botswana. The potential of upcycling to reconfigure waste into valuable resources aligns with the principles of the circular economy, offering a beacon of hope for waste reduction, conservation of resources and the minimization of environmental impact. This study's insights into students' attitudes and perceptions towards upcycling highlight the capacity of design education to shape sustainable mindsets. By immersing students in the ethos of upcycling, industrial design programmes can empower a new generation of designers who view creativity as harmonizing human creativity with ecological balance. The potential long-term impact of this transformation is profound, extending beyond classrooms to permeate the very fabric of industrial practices. The study's emphasis on successful upcycling integration underscores the interconnectedness of stakeholders in the journey towards sustainable design. The roles of government agencies, environmental organizations, industry partners, and academia converge to weave a tapestry of possibilities. As participants noted, these collaborative efforts can shape policy frameworks, provide necessary resources, and amplify the reach of upcycling initiatives.

As the study draws to a close, it offers a series of recommendations that emerge from the nuanced insights gathered from participants in both qualitative discussions and quantitative surveys. These recommendations, ranging from curriculum development to advocacy efforts, provide a roadmap for institutions, educators, policymakers, and practitioners to embark on a transformative journey that integrates upcycling into the very fabric of industrial design education and practice. In conclusion, the integration of upcycling principles in the industrial design programme heralds the dawn of a new era – one that bridges design creativity with environmental stewardship and economic resilience. As Botswana navigates its path towards sustainable development, the marriage of upcycling and design education emerges as a beacon of hope. This path not only promises to reshape industrial landscapes but also fosters a harmonious coexistence between human innovation and the natural world.

References

Andersen, M. M. (2007), 'An introductory note on the environmental economics of the circular economy', *Sustainability Science*, 2(1): 133–140.
Baldwin, J., Bento, R. F. and Bower, M. (2018), 'A study of factors influencing student engagement in sustainability in higher education', *Journal of Cleaner Production*, 171: 1181–1191.
Bennett, G. (2019), 'Upcycling and the circular economy: A review of the literature and future research opportunities', *Journal of Cleaner Production*, 228: 1580–1592.
Braun, V. and Clarke, V. (2006), 'Using thematic analysis in psychology', *Qualitative Research in Psychology*, 3(2): 77–101.
Brundtland, G. H. (1987), *Our common future: Report of the 1987 World Commission on Environment and Development*, United Nations.
Chinda, T. and Suri, R. (2020), 'Implementing the sustainable development goals (SDGs) in engineering education: A review of pedagogical approaches', *Sustainability*, 12(22): 9517.
Creswell, J. W. and Creswell, J. D. (2018), *Research design: Qualitative, quantitative, and mixed methods approaches*, London: Sage Publications.
DEA (2015), 'Fifth National Report to the Convention on Biological Diversity', Department of Environmental Affairs, Republic of Botswana, Available at: https://www.cbd.int/doc/world/bw/bw-nr-05-en.pdf.
Fletcher, K. (2014), *Sustainable fashion and textiles: Design journeys*, London: Routledge.
Geissdoerfer, M., Savaget, P., Bocken, N. M. and Hultink, E. J. (2017), 'The circular economy – A new sustainability paradigm?', *Journal of Cleaner Production*, 143: 757–768.
Ghisellini, P., Cialani, C. and Ulgiati, S. (2016), 'A review on circular economy: The expected transition to a balanced interplay of environmental and economic systems', *Journal of Cleaner Production*, 114: 11–32.
Hoornweg, D. and Bhada-Tata, P. (2012), 'What a waste: A global review of solid waste management. Urban development series'. World Bank. Available at: https://openknowledge.worldbank.org/entities/publication/1a464650-9d7a-58bb-b0ea-33ac4cd1f73c.
Lange, S. and Rennings, K. (2017), 'Sustainability-oriented innovation of SMEs: A systematic review', *Journal of Cleaner Production*, 162: 569–590.
Palinkas, L. A., Horwitz, S. M., Green, C. A., Wisdom, J. P., Duan, N. and Hoagwood, K. (2015), 'Purposeful sampling for qualitative data collection and analysis in mixed

method implementation research', *Administration and Policy in Mental Health and Mental Health Services Research*, 42(5): 533–544.
Sonneveld, U. and Wiersma, W. (2011), 'Upcycling in industrial design education', *Proceedings of E&PDE 2011, 13th International Conference on Engineering and Product Design Education*, London.
United Nations (2015), 'Transforming our world: The 2030 agenda for sustainable development', *United Nations*, Available at: https://sdgs.un.org/2030agenda.v

Conclusion

Opportunities and Action Needed

Richie Moalosi and Kyungeun Sung

This concluding chapter is based on the thematic areas that emerged from the chapters that the authors submitted. These include fashion and textiles, engineering and the built environment, design and business, people and the community and design education.

Upcycling in Fashion and Textiles

The journey of upcycling in fashion and textiles has been both transformative and revolutionary, redefining how the industry approaches sustainability, creativity and consumer engagement. As we conclude this exploration into the world of upcycled fashion, it is essential to reflect on the key themes, achievements and future directions that will continue to shape this dynamic field. Upcycling has fundamentally altered the perception of waste in the fashion industry. By viewing discarded materials as valuable resources rather than refuse, designers have unlocked a new realm of creative possibilities (Cassar, 2024). This shift in mindset has led to innovative design practices that prioritize resource efficiency and environmental stewardship. The success of upcycled fashion demonstrates that sustainability and style are not mutually exclusive but can coexist harmoniously.

One of the most significant contributions of upcycling is its positive environmental impact. By diverting textile waste from landfill and reducing the demand for virgin materials, upcycling helps mitigate the fashion industry's substantial ecological footprint. The reduction in greenhouse gas emissions, water consumption and chemical usage associated with upcycled products underscores the critical role this practice plays in promoting a more sustainable and responsible industry (Bhatt, 2020). Upcycling also offers promising economic and social benefits. It creates opportunities for local artisans, small businesses and marginalized communities to participate in the fashion industry. In some unusual cases, fashion upcycling could also become a medium for transforming people's negative experiences into empowerment (stitched refugee stories in Chapter 4). By fostering a culture of craftsmanship and supporting ethical labour practices, upcycling contributes to a more inclusive and equitable fashion landscape (e.g., Kenyan fashion designers' practice in Chapter 3 and fashion upcycling practices in Harare, Zimbabwe in Chapter 5). Moreover, the unique nature of upcycled products, often characterized by limited editions and one-of-a-kind pieces, appeals to consumers seeking authenticity and individuality in their fashion choices. In addition, the fashion upcycling design that incorporates 'emotional design' (Chapter 2) might have a higher chance of attracting more consumers.

The rise of upcycling has been accompanied by a growing consumer awareness of the environmental and social impacts of their purchasing decisions. As consumers become more conscious of the sustainability credentials of the products they buy, there is an increasing demand for transparency and accountability from fashion brands and designers. This shift in consumer behaviour drives the industry toward more sustainable practices, encouraging brands to adopt upcycling and other eco-friendly initiatives. A report published in 2020 by McKinsey & Company (2020) indicates that 67 per cent of consumers consider sustainable materials when making purchases in the fashion sector, while 63 per cent prioritize brands that exhibit a commitment to environmental responsibility. This evolution in consumer behaviour is compelling brands to adopt upcycling practices to satisfy the increasing demand for eco-friendly products. Perhaps an accessible mobile app to make clothing sustainable (Chapter 6) could be offered by major fashion companies in the future.

Despite its many benefits, upcycling in fashion and textiles faces several challenges. Scalability, quality control, cost, consumer perception and the availability of suitable materials are significant hurdles that need to be addressed. Upcycled products typically necessitate a greater investment of labour and time, rendering them more costly than items produced through mass manufacturing. However, as advancements in technology occur and consumer demand continues to rise, it is anticipated that these challenges will gradually be mitigated (Henninger et al., 2016). These challenges also present opportunities for innovation. Advances in technology, such as automated sorting systems, digital fabrication, 3D printing and AI-driven material sorting, hold the potential to overcome these barriers and enhance the efficiency and viability of upcycling practices. Researchers are also exploring chemical recycling methods to break down blended fabrics into reusable fibres, which could revolutionize upcycling practices (Ellen MacArthur Foundation, 2021).

Collaborations among fashion brands, textile manufacturers and waste management firms are driving innovation in the field of upcycling. Industry-wide initiatives and partnerships can drive systemic change and create a supportive ecosystem for upcycled fashion. For example, brands such as Patagonia and Eileen Fisher have established partnerships with textile recycling firms to develop upcycled collections (Fletcher, 2014). These collaborations are crucial for enhancing upcycling initiatives and fostering a more sustainable circular fashion system. Policy interventions, such as extended producer responsibility and incentives for sustainable practices, can further promote the adoption of upcycling and other circular economy principles. Designers should strive to embrace the core principles of sustainable fashion, including: (i) supply chain transparency, (ii) ethical labour practices for production/manufacturing workers, (iii) collaborative consumption, (iv) circularity, recycling and upcycling and (v) production methods with a minimal impact on the environment (Bardon, 2024). Education and advocacy are critical components in advancing sustainability within the fashion industry (ibid.). Institutions such as the London College of Fashion and Parsons School of Design are integrating upcycling into their academic programmes, thereby equipping the forthcoming generation of designers with the skills necessary to develop sustainable fashion practices (Fletcher and Tham, 2015). As designers in the fashion sector become more knowledgeable and aware of sustainability challenges, they are better positioned to effect meaningful change.

Upcycling in Engineering and the Built Environment

Upcycling in engineering and the built environment represents a fundamental shift in the perception and use of waste materials and is increasingly recognized as a sustainable practice within these sectors. Rather than being discarded, materials are seen as valuable resources that can be repurposed and reintegrated into new projects, reducing waste and improving cost-effectiveness (e.g., upcycling carbon-fibre reinforced polymer matrix composites in Chapter 7 and unlocking the value from coffee waste in Chapter 10). The construction industry contributes significantly to global waste, accounting for about 30 per cent of all waste worldwide (World Bank, 2018), where upcycling could be a viable solution (integrating upcycling into urban infrastructure and the built environment in Chapter 9). For instance, reclaimed materials such as steel, concrete and wood are being reused in new building projects (Adams et al., 2017). Economically, upcycling can lead to significant cost savings in material procurement and waste disposal while also stimulating new markets for recycled and upcycled materials (e.g., upcycling waste solar photovoltaic panels in Chapter 8). Recent advancements in materials science and engineering are paving the way for the development of innovative upcycled materials. Researchers are investigating the use of recycled plastic waste to create lightweight and durable construction materials, including bricks, tiles and insulation panels (Hopewell et al., 2009). Furthermore, 3D printing technology is being employed to manufacture upcycled building components, providing enhanced precision and customization while simultaneously reducing material waste (Ford and Despeisse, 2016). Upcycling contributes to the resilience and adaptability of buildings and infrastructure. By incorporating reclaimed materials, structures can be designed to be more adaptable to changing needs and conditions. This flexibility is crucial in the face of climate change, urbanization and other dynamic factors impacting the built environment. Upcycled materials often possess unique characteristics that can enhance the durability and aesthetic value of projects, further supporting long-term resilience. Urban environments are progressively integrating upcycled materials into infrastructure projects. For instance, reclaimed asphalt pavement is utilized in road construction, while upcycled shipping containers are being transformed into modular housing and commercial spaces (Cheng et al., 2020). These approaches not only minimize waste but also enhance the aesthetic and functional diversity of urban landscapes.

Despite its significant potential, upcycling within the fields of engineering and the built environment has encountered several challenges, including scalability, cost considerations and regulatory obstacles. For instance, the processing of upcycled materials may prove to be labour-intensive and more costly compared to traditional materials. However, it is anticipated that advancements in technology and the realization of economies of scale will mitigate these challenges over time (Ghisellini et al., 2016). Innovations in material science, digital fabrication and design software are making it easier to identify, process and integrate reclaimed materials into new projects. These technologies also enable more precise and efficient construction practices, enhancing the feasibility and attractiveness of upcycling as a mainstream approach. Policy and regulatory frameworks are essential in promoting and supporting upcycling initiatives. Governments and regulatory bodies can implement incentives, standards and guidelines that encourage the use of upcycled materials in construction and engineering projects. Extended producer responsibility programmes, green building certifications and circular economy policies are examples of mechanisms that can drive the adoption of upcycling practices. Governments and international organizations increasingly acknowledge the significance of upcycling in the pursuit of sustainability objectives. For example, the European Union's Circular Economy Action Plan advocates for the integration of recycled and upcycled materials in construction and infrastructure initiatives (European Commission, 2020). Likewise, the United Nations Sustainable Development Goals (SDGs) underscore the necessity for sustainable urban environments and communities, thereby promoting the implementation of upcycling practices (United Nations, 2015).

Upcycling in Design and Business

The practice of upcycling in design and business has emerged as a revolutionary approach that intertwines sustainability, creativity and economic viability. Upcycling redefines the concept of value in both design and business; instead of discarding products that have reached the end of their initial life cycle, upcycling leverages these to create new products/materials of higher quality or value than the original (e.g., furniture, lighting and bags made through upcycling in Chapter 11 and designing with waste in Hong Kong in Chapter 13). Upcycling promotes the advancement of sustainable business models that emphasize resource efficiency and the reduction of waste, whether organic (Chapter 14) or residual waste (e.g., recyclables, non-recyclables and hazardous waste). Organizations are incorporating upcycling into their operations to address consumer demand for environmentally friendly products and to align with their corporate social responsibility objectives. A notable example is Patagonia's 'Worn Wear' programme, which focuses on repairing and upcycling used clothing, thereby extending the lifecycle of its products and mitigating environmental impact (Patagonia, 2021). In a similar vein, start-ups such as Rothy's, manufacture shoes and bags from recycled plastic bottles, effectively demonstrating the profitability associated with upcycled products (Rothy's, 2023).

Upcycling presents numerous economic opportunities for businesses. It opens new markets and revenue streams by creating unique, value-added products that appeal to environmentally conscious consumers. Companies that embrace upcycling can differentiate themselves in the marketplace, fostering brand loyalty and attracting a growing segment of eco-conscious customers. Consumer preferences are increasingly shifting toward sustainable and ethical products, resulting in a burgeoning market for upcycled goods. According to a 2021 report by Nielsen (2021), 66 per cent of global consumers expressed a willingness to pay a premium for products offered by sustainable brands. This trend is motivating businesses to implement upcycling practices as a means of differentiation and to better cater to environmentally conscious consumers. Moreover, upcycling encourages innovation, pushing designers and businesses to think creatively about material usage and product design. For businesses, upcycling offers a powerful way to enhance brand identity and build stronger connections with consumers. By adopting sustainable practices and creating upcycled products, companies can communicate their commitment to environmental stewardship and social responsibility. This resonates with consumers who prioritize sustainability, leading to increased brand loyalty and customer engagement. Designers are harnessing the power of upcycling to produce unique, high-quality products that merge sustainability with creativity. For instance, Dutch designer Piet Hein Eek is acclaimed for his furniture crafted from reclaimed wood and scrap materials, highlighting the aesthetic possibilities of upcycling (Eek, 2023). Furthermore, advancements in digital design tools and 3D printing are empowering designers to explore innovative applications of upcycled materials (Ford and Despeisse, 2016).

Upcycling businesses foster community involvement and support local economies. By sourcing materials locally and involving communities in the upcycling process, businesses can create jobs, develop skills and promote economic growth. Upcycling projects often engage local artisans and craftsmen, preserving traditional skills and promoting cultural heritage. Despite its many advantages, upcycling businesses face challenges such as low consumer awareness, time-consuming processes, lack of facilities/equipment, difficulties in sourcing consistent, high-quality materials and legal issues (Chapter 12). However, these challenges present opportunities for innovation and collaboration. The future of upcycling in design and business depends on collaborative networks that connect designers, manufacturers and waste management systems. Platforms such as the Ellen MacArthur Foundation's Circular Design Network promote collaboration and knowledge-sharing to enhance upcycling initiatives (Ellen MacArthur Foundation, 2021). Advances in technology, such as improved material processing techniques and digital design tools, can help overcome these barriers. Additionally, education and awareness campaigns can shift consumer perceptions, highlighting the benefits and desirability of upcycled products. Supportive policies and regulations are crucial for the growth of upcycling in design and business. Governments and regulatory bodies can encourage upcycling through incentives, tax breaks and grants for businesses adopting sustainable practices. Implementing standards and certifications for upcycled products can also boost consumer confidence and promote market acceptance.

Upcycling on People and the Community

The practice of upcycling has brought about significant changes in communities worldwide, fostering environmental stewardship, economic resilience and social cohesion (see examples of community participation projects in Thailand in Chapter 16 and community empowerment project in Zimbabwe in Chapter 18). Upcycling has empowered individuals and communities by turning waste into resources and fostering a culture of creativity and self-reliance (see the good practice in Brazil in Chapter 17). The object of upcycling can also be extended to people: both unemployed and retired people can be upcycled into useful labour in society (Chapter 15). Upcycling promotes a sense of agency and empowerment, allowing individuals to contribute meaningfully to their communities and the environment. Upcycling initiatives serve to empower marginalized communities by providing avenues for skill development, employment and entrepreneurship. For example, organizations such as Trashy Bags in Ghana employ local workers to convert plastic waste into bags and accessories, thereby generating jobs and addressing environmental pollution (Trashy Bags, 2023). These initiatives not only address waste management challenges but also foster social inclusion and strengthen economic resilience. By supporting local upcycling initiatives, communities can develop more resilient economies that are less dependent on external resources.

Community upcycling promotes environmental education and awareness, motivating individuals and communities to embrace sustainable practices. An increasing number of schools and community centres are integrating upcycling workshops into their curricula, thereby instructing participants on the creative repurposing of waste materials. These initiatives foster a culture of sustainability and resourcefulness, particularly among younger generations (UNESCO, 2017). The initiatives also often bring together diverse groups of people, encouraging collaboration and shared purpose. These projects can strengthen community bonds, promote inclusivity and foster a sense of belonging. By working together on upcycling initiatives, community members can build stronger, more connected communities. Community upcycling plays a significant role in preserving cultural heritage by integrating traditional crafts and materials into contemporary designs. For instance, artisans in India upcycle saris and various textiles to create new products, effectively blending cultural traditions with modern sustainability practices (Fletcher and Tham, 2015). This combination of heritage and innovation not only fosters cultural pride but also encourages creative advancement. By embracing upcycling, we can create communities that are not only more sustainable but also more vibrant and resilient. The future of upcycling for individuals and the community is encouraging and its ongoing development will be essential in promoting a more sustainable and inclusive world.

Upcycling on Design Education

The integration of upcycling into design education has ushered in a life-changing era of sustainability, creativity and innovation, shaping the next generation of designers who are equipped to address global sustainability challenges (e.g., the case of the University of Botswana in Chapters 19 and 22). This practice encourages students and educators to rethink the lifecycle of materials and to view waste as an opportunity for creative expression and problem-solving. Upcycling is increasingly being integrated into design curricula and has redefined design education by incorporating sustainability and circularity as a core principle. For example, several institutions, including the University of Botswana, the University of Cincinnati, the Royal College of Art and Colgate University have established programmes that focus on sustainable design practices and circular economy, particularly upcycling (Fletcher and Tham, 2015). These educational initiatives underscore the significance of resource efficiency, waste minimization and ethical production methods. By challenging students to create value from discarded materials, upcycling pushes the boundaries of traditional design thinking. Students learn to experiment with unconventional materials, develop novel solutions and embrace the aesthetic and functional potential of upcycled products (see examples in Chapters 20 and 21). This creative approach not only enhances their design skills but also instils a mindset of resourcefulness and adaptability. This education empowers them to become advocates for sustainability in their professional careers. Upcycling serves as a bridge between theoretical knowledge and practical application in design education. Students engage in hands-on projects that require them to apply theoretical concepts to real-world challenges. This experiential learning approach enhances their soft skills, such as problem-solving skills, critical thinking and the ability to work with diverse materials. By working on upcycling projects, students develop a practical understanding of sustainable design principles that can be applied in their future careers. In addition, students from different disciplines can work together on upcycling projects, combining their expertise to develop comprehensive solutions. This collaborative approach fosters a deeper understanding of the interconnectedness of design, sustainability and other fields, preparing students to work effectively in diverse, multidisciplinary teams.

Introducing upcycling into design education, despite its importance in addressing pressing environmental issues, comes with its own set of challenges, such as limited access to suitable materials, variability in material quality and the need for specialized tools and knowledge. However, these challenges also present opportunities for partnerships and collaborations with industry to source diverse materials, incorporate new technologies and provide practical experience for students. For example, partnerships with organizations such as Patagonia and IKEA enable students to undertake upcycling projects that address genuine waste streams and consumer needs (Ellen MacArthur Foundation, 2021). These collaborations effectively bridge the gap between academia and industry, equipping students for careers in sustainable design. Supportive policies and institutional frameworks are crucial for the successful integration of upcycling into design education. Educational institutions can promote upcycling through dedicated programmes, funding for sustainable projects and partnerships with industry and community organizations. By embedding upcycling into their sustainability strategies, institutions can create an enabling environment for students to engage in meaningful and impactful design work.

Summary

In *Upcycling Research: Global Perspectives on Product, Fashion and Theory*, we have explored the multifaceted impact of upcycling across various sectors. This comprehensive exploration underscores the transformative potential of upcycling in fostering sustainability, creativity and economic resilience. From fashion to design education, upcycling challenges traditional paradigms, turning waste into valuable resources and inspiring innovation. It promotes environmental stewardship, empowers communities and prepares future generations for a more sustainable and equitable world. As we conclude this book, it is evident that upcycling is not merely a trend or façade, but a powerful and sustainable movement – one that reimagines how we engage with resources and design our surroundings, ultimately shaping a more sustainable future across all areas of society.

This book offers several calls to action, especially to policymakers, educators, industry leaders and consumers, all of whom have roles to play in advancing the upcycling movement. Policies that support sustainable practices, educational initiatives that raise awareness of the benefits of upcycling and consumer choices that prioritize upcycled products can collectively drive meaningful change. By embracing upcycling, we can not only reduce waste and conserve resources but also promote a more sustainable and equitable world. The journey of upcycling is one of continual learning and adaptation. As we move forward, we must remain committed to the principles of sustainability, creativity and community that underpin this transformative movement.

References

A (2020), 'Circular Economy Action Plan'. Available at: https://ec.europa.eu/.

Ford, S. and Despeisse, M. (2016), 'Additive manufacturing and sustainability: An exploratory study of the advantages and challenges', *Journal of Cleaner Production*, 137, 1573–1587.

Fletcher, K. (2014), '*Sustainable Fashion and Textiles: Design Journeys*'. Abingdon: Routledge.
Fletcher, K. and Tham, M. (2015), '*Routledge Handbook of Sustainability and Fashion*'. Abingdon: Routledge.
Ghisellini, P., Cialani, C. and Ulgiati, S. (2016), 'A review on circular economy: The expected transition to a balanced interplay of environmental and economic systems', *Journal of Cleaner Production*, 114, 11–32.
Henninger, C.E., Alevizou, P.J. and Oates, C.J. (2016), 'What is sustainable fashion?' *Journal of Fashion Marketing and Management*, 20(4), 400–416.
Hopewell, J., Dvorak, R. and Kosior, E. (2009), 'Plastics recycling: Challenges and opportunities', *Philosophical Transactions of the Royal Society B: Biological Sciences*, 364(1526), 2115–2126.
McKinsey and Company (2020), '*The State of Fashion 2020*'. Available at: https://www.mckinsey.com/.
Nielsen (2021), 'Global Sustainability Report'. Available at: https://www.nielsen.com/.
Patagonia (2021), '*Worn Wear Program*'. Available at: https://wornwear.patagonia.com/.
Rothy's (2023), '*Sustainable Products Made from Recycled Materials*'. Available at: https://www.rothys.com/.
Trashy Bags (2023), '*About Us*'. Available at: https://www.trashybags.org/.
UNESCO (2017), 'Education for Sustainable Development Goals'. Available at: https://en.unesco.org/
United Nations (2015), 'Sustainable Development Goals'. Available at: https://www.un.org/sustainabledevelopment/.
World Bank (2018), '*Solid Waste Management*'. Available at: https://www.worldbank.org/.

INDEX

Numbers in **bold** indicate figures.